# UNDERSTANDING OUR ATMOSPHERIC ENVIRONMENT

*View from ATS 3 Satellite stationed over the equator at 63°W, taken at local noon (1612 GMT), September 30, 1970. [Courtesy of National Environmental Satellite Center, NOAA.]*

# UNDERSTANDING OUR ATMOSPHERIC ENVIRONMENT

**Morris Neiburger**
University of California, Los Angeles

**James G. Edinger**
University of California, Los Angeles

**William D. Bonner**
National Oceanic and Atmospheric Administration

W. H. FREEMAN AND COMPANY    San Francisco

Cover design from a photograph by the National Oceanic and Atmospheric Administration.

*Library of Congress Cataloging in Publication Data*

Neiburger, Morris.
Understanding our atmospheric environment.

1. Meteorology. 2. Weather. I. Edinger, James G., joint author. II. Bonner, William D., joint author. III. Title.
QC861.2.N45     1973     551.5     72–4753
ISBN 0–7167–0257–6

*Printed in the United States of America*

*International Standard Book Number: 0-7167-0257-6*

*3 4 5 6 7 8 9*

# CONTENTS

**Preface    ix**

**1    The Drama of the Weather    1**

1.1 The excitement of meteorology    1
1.2 Characters in the drama. Clouds    3
1.3 Weather systems and the weather map    10
1.4 New tools for meterorology    17
1.5 The role of mathematics in science    21

**2    The Composition and Thermal Structure of the Atmosphere    23**

2.1 Nature and composition of the atmosphere    23
2.2 Vertical structure of the atmosphere    26
2.3 Latitudinal and seasonal variations of temperature at various heights    32

**3    Radiation Through the Atmosphere    39**

3.1 The sun and the earth    39
3.2 The nature and laws of electromagnetic radiation    40
3.3 Energy, heat, and work    45
3.4 Solar radiation    47
3.5 Insolation at the earth's surface    50
3.6 Depletion of solar radiation by the atmosphere    56
3.7 Transfer of terrestrial radiation through the atmosphere    60

**4    The Gas Laws. Heat and Temperature Changes    66**

4.1 The behavior of gases    66
4.2 Variation of pressure with height    72
4.3 Heat and temperature change    74
4.4 Thermodynamic diagrams. Process curves and sounding curves    77
4.5 Atmospheric stability    79

## 5 Motions in the Atmosphere. Small-Scale Circulations 86

5.1 Air in motion 86
5.2 Newton's laws of motion 88
5.3 Forces in a fluid 89
5.4 Causes of horizontal pressure variations 92

## 6 Large-Scale Motions 99

6.1 The effect of the earth's rotation 99
6.2 Quantitative expression for the Coriolis force 101
6.3 The geostrophic wind 104
6.4 Curved flow. The gradient wind 106
6.5 The effect of friction 109
6.6 Variation of pressure gradients with height. The thermal wind 111

## 7 The General Circulation of the Atmosphere 115

7.1 Deflection of meridional circulations 115
7.2 Radiation balance 116
7.3 Hadley cells and angular momentum 117
7.4 Transport of heat and momentum by waves and vortices 120
7.5 Experiments on the general circulation 120
7.6 The observed average-wind distribution 122
7.7 The observed mean pressure distribution 125

## 8 Water in the Atmosphere. Condensation and Precipitation 131

8.1 Humidity variables 131
8.2 Condensation 135
8.3 The saturation adiabatic process 139
8.4 Stability of cloudy air 142
8.5 Clouds and precipitation 143
8.6 Forms of precipitation 148

## 9 Convection. Cumulus Clouds, Thunderstorms, and Tornadoes 153

9.1 The development of convective clouds 153
9.2 Influence of the environment. Entrainment 155
9.3 Processes that produce changes in stability 159
9.4 Relation of convective activity to flow patterns 162
9.5 Tornadoes 165

## 10 Air Masses 168

10.1 The nature and classification of air masses 168
10.2 Properties of air masses 169
10.3 Transformation of air masses 170
10.4 Air-mass weather 174

## 11 Fronts and Cyclones    177

11.1 Introduction    177
11.2 Fronts    178
11.3 Frontal waves. The wave cyclone    181
11.4 The life cycle of a cyclone. The occlusion process    184
11.5 Upper-level flow in relation to wave cyclones. Waves in the westerlies.
    The jet stream    188
11.6 Cyclogenesis    191

## 12 Weather Systems in the Tropics    194

12.1 The atmospheric circulation at low latitudes    194
12.2 Characteristics of the trade-wind region    196
12.3 Waves in the easterlies    197
12.4 Tropical cyclones, hurricanes, and typhoons    198

## 13 Weather Forecasting    208

13.1 Methods of weather prediction    208
13.2 Predictability of the weather. Accuracy of weather forecasts    209
13.3 Analysis of the present weather    213
13.4 Numerical weather prediction    220
13.5 Special forecasts    231
13.6 Extended-range forecasting    234
13.7 Programs for the improvement of forecasting    235

## 14 Air Pollution    238

14.1 Types and sources of air pollution    238
14.2 Factors affecting the concentration of air pollution    241
14.3 Evaluation of air-pollution potential    245
14.4 Effects of air pollution    250
14.5 Control of air pollution    252

## 15 Man's Influence on the Atmosphere. Inadvertent and Intentional Modification    256

15.1 Effects of human activities. The weather in cities    256
15.2 Influences on worldwide weather    258
15.3 Intentional weather modification    261

**Appendix**   A. Units of measurement and conversion factors    271
           B. Numerical constants    273
           C. Greek alphabet    274
           D. Weather-map symbols    274

**Index**    283

# PREFACE

Interest in atmospheric science is a pervasive and continuing human attitude. The old saying "Everybody talks about the weather . . . ." is a truism that originates from this interest. The extension of our observational resources to higher and higher altitudes via rockets and satellites and the increasing heights at which commercial airplanes are operated have added a third dimension to the interest. For instance, we now talk about the "weather" in terms of the effects of SST flights on ozone in the stratosphere and the effects of CAT—clear air turbulence—on high-level airplane flights. The concluding clause of the saying, that "nobody does anything about it," never completely true, is contradicted by the expanding activity in the field of weather modification.

This book attempts to answer some of the questions that arise from this interest. Its emphasis is on *understanding*, that is, on providing the student with the physical explanations of atmospheric phenomena. As such, its purpose is twofold: to give the student a deeper appreciation of this part of the physical universe, which he experiences so intimately and continuously, and to impart an awareness of some of the laws of physics and the way in which they are applied.

In the process of communicating this understanding, a considerable amount of descriptive information about the structure of the atmosphere and the processes that go on in it is presented. The temperature distribution, circulation patterns, and structure of storms are described and explained. However, the scope of meteorology is too extensive for all phases to be included in a textbook intended for a one-quarter or one-semester course. Most of them are referred to at least in passing, but only those central to the weather as experienced by an observer at the ground or in flight on an airplane are dealt with in detail.

The book is intended for students who are majoring in other subjects than a physical science. Although it contains mathematics at a level usually required for admission to college, for most of the concepts it is not necessary to follow the mathematics.

The first eleven chapters are based on mimeographed notes that were written several years ago. They were revised and published, with the addition of the twelfth chapter, as a preliminary edition in 1971. Based on the experience of using the notes and the preliminary edition and on comments from others concerning them, the text has been modified further and expanded. The present book contains more material than can easily be accommodated in a quarter or even a semester. Although some of the material in the first twelve chapters can be omitted (for instance, the occasional interpolation of historical accounts of the development of meteorological understanding), these chapters constitute the essence of the subject. On the other hand, the last three chapters contain some of the more interesting applications of meteorology, and instructors may wish to minimize the discussion of earlier chapters in order to include them.

For the mimeographed notes, the original draft of chapters 5 and 9 was written by Professor Edinger, and that of chapters 6, 7, 10, and 11 was written by Dr. Bonner. The undersigned wrote the remainder of the original draft and revised the entire manuscript. The text has undergone three revisions: one for the mimeographed version, one for the preliminary edition, and one for the present publication. At these various stages we benefited from suggestions made by several people who used or reviewed portions of the original notes or the preliminary edition. Professors A. Arakawa, M. Yanai, and R. M. Thorne, of the Department of Meteorology of the University of California, Los Angeles, and Professor Arnold Court, of the Department of Geography of California State University, Northridge, examined portions of the notes or the manuscript of the preliminary edition. In addition, the preliminary edition was class-tested or critically examined by

Professor Phillip Bacon
  *University of Houston*

Professor Harold C. Bulk
  *California State University,
  Hayward*

Professor Eugene Chermack
  *State University of New York,
  Oswego*

Professor L. Glen Cobb
  *University of Northern
  Colorado*

Professor Harry C. Coffin
  *California State University,
  Los Angeles*

Professor Dennis M. Driscoll
  *Texas A&M University*

Professor G. Thomas Foggin III
  *University of Montana*

Professor Gary Hannes
  *California State University,
  Fullerton*

Professor Donald R. Haragan
  *Texas Tech University*

Professor James M. Havens
  *University of Rhode Island*

Professor Donald W. Lea
  *Southeastern Louisiana College*

Professor Paul E. Lydolph
  *University of Wisconsin,
  Milwaukee*

Professor J. H. McCaughey
*Queen's University*

Professor John E. Oliver
*Columbia University*

Professor Robert Phillips
*University of Wisconsin,
Platteville*

Professor Gerald L. Potter
*California State University,
Hayward*

Professor John N. Rayner
*The Ohio State University*

Professor Richard G. Reider
*University of Wyoming*

Professor Barry Saltzman
*Yale University*

Professor Allen E. Staver
*Northern Illinois University*

Professor John W. Stewart
*University of Virginia*

Professor N. Sundararaman
*California State University,
Northridge*

Professor Neil B. A. Trivett
*Indiana University*

Professor Jack R. Villmow
*Northern Illinois University*

Professor D.M. Welch
*The University of Winnipeg*

Professor James D. Wilson
*East Tennessee State University*

Professor Richard E. Witmer
*East Tennessee State University*

Professor Constantine Zois
*Newark State College*

A number of the above made valuable suggestions that influenced the preparation of the final manuscript. Especially helpful were the detailed comments of Professor James M. Havens, Professor Allen E. Staver, Professor John W. Stewart, and Professor Narasimhan Sundararaman.

I wish to acknowlege with thanks the permission by various copyright holders to reproduce figures and illustrations. The particular source of each figure is indicated in its legend.

Mrs. A. C. Rizos assisted with the preparation of the manuscript and typed the copy for the various versions. I want to express my thanks to her for her advice and good judgment as well as her careful work.

Morris Neiburger

November, 1972

# ONE

## The Drama
## of the
## Weather

### 1.1   The excitement of meteorology

There are many reasons for studying meteorology. To be sure, one of them is to fulfill a curricular requirement. Among the various alternatives for satisfying science requirements a course in meteorology seems to many students more attractive or less painful than others. We hope this expectation is fulfilled. But we also hope that studying meteorology will enhance the student's ability to cope with his or her environment and derive new pleasures from it.

The objective of science is to observe and explain the physical world —the world of nature. In meteorology, it is the phenomena occurring in the atmosphere that are to be explained. The atmosphere is the most intimate part of our environment, constantly surrounding us and even entering within us with every breath we take. To understand this part of our environment is to be better able to relate to it and to make decisions regarding it. These decisions include the daily personal ones: What kind of clothes shall I wear? Shall I carry an umbrella? Shall I go surfing or skiing? They also include political decisions regarding the environment: Shall we build the SST? Shall we support experiments to increase rain? What shall we do about smog?

In addition to enabling the student better to deal with his individual environmental problems and his responsibilities as a citizen, studying meteorology will enable the student to share the sense of excitement of penetrating some of the mysteries of nature and the pleasure associated with appreciating the logic, order, and beauty in the laws that it obeys. It is this excitement that motivates researchers in the sciences, and indeed in all fields. The students who plan to specialize in sociology, psychology, political science, or economics are motivated by an interest in the behavior of man and society. The zoologist, botanist, bacteriologist, and biochemist are spurred by the same spirit in their attempt to understand living organisms. And the physical scientists, among whom are numbered the meteorologists, are motivated by the same eagerness to understand the "hows and whys" of the inanimate world about us. To partake of the gratification that comes from this understanding is the main reason why the liberal arts student should take courses in physical science.

Physicists and chemists are mostly concerned with things that are small, such as molecules, atoms, or even subatomic particles such as neutrons, electrons, mesons, and neutrinos. Astronomers are concerned with things that are very large, such as stars and galaxies. Meteorologists are concerned with things that range from microscopic (e.g., condensation nuclei) to earth-sized (general wind systems). Are meteorological phenomena as dramatic as splitting the atom or as awe-inspiring as the expanding galaxies? The drama of a developing thunderstorm, the awesome fury of a hurricane, or the mystery surrounding the growth of a rain drop or snow flake all excite an equally great sense of wonder in those who become aware of the processes and forces that produce them.

The beauty in meteorology lies not only in abstract processes and laws of forces. The myriad cloud forms, the manifold colors of the sky at sunset or its blueness at noon, auroral displays at night, glistening dew on the meadow at daybreak, the infinitely varied forms of snow flakes—all have intrinsic beauty that is augmented by an understanding of their causes.

To this beauty is added the excitement and adventure of never-ending change and development: a drama with plot within plot, a poem with intricate rhythm and rhyme, a novel in which characters grow and interact. Indeed, the titles of many literary works—Shakespeare's "Tempest," Conrad's "Typhoon," George Stewart's "Storm," Norman Douglas's "South Wind," and countless others—reflect their authors' awareness of the colorful and dramatic qualities of meteorological phenomena.

Many of these phenomena are familiar. From common experience everyone is aware in a qualitative way of the daily and annual course of temperature (warm at noon and in summer, cold at night and in winter) and the irregular alternation of clear and cloudy days, or dry days and days with rain or snow, and of sparklingly transparent air and haze or fog. Almost no one has escaped the effects of severe weather: drenching, flood-producing rains, severe thunderstorms, traffic-paralyzing snows, or destructive winds. News accounts of these disasters have made them familiar to everyone, and television has brought vivid pictures of those occurring in remote places into our homes.

Other meteorological phenomena, just as interesting, are not part of our daily conscious experience. The variations of atmospheric pressure, for instance, are as important as those of temperature and humidity, but prior to the invention of the barometer no one was aware of them. Pressure variations are connected with moving storm systems, to such an extent that the storms are usually identified with the associated low pressure centers. The pressure has diurnal and semidiurnal oscillations that are conspicuous in tropical regions but become obscured by the variations due to moving storms in higher latitudes. The serene rhythm of fair weather and the dramatic conflicts of severe storms are represented as well in the pressure variations we do not feel as in the temperature variations, the buffeting winds, the tumultuous clouds, and the torrential rains that affect our tactile, visual, and even our auditory senses.

There are many other atmospheric phenomena that are not part of our direct experience. Some, like the ionosphere, which reflects radio waves and makes possible long distance radio reception, have indirect influences on our lives or activities. Others, like the airglow (a chemiluminescence occurring in the upper atmosphere), have no obvious effects on man and are not noticeable to the casual observer. These phenomena, too, are interesting to understand.

## 1.2 Characters in the drama. Clouds

Among the most interesting characters in the weather drama are the clouds. Consider, first, the typical sequence of clouds on a quiet summer day in the south-central United States. The early morning is clear, but as the day progresses and the temperature rises scattered small heap clouds begin to form. These heap clouds are named *cumulus*. They have distinct flat bottoms all at about the same height. The tops are rounded, in the form of dense mounds or domes, and are brilliantly white in the sunlight. With time the tops grow higher and higher, the growth taking the form of protuberances or towers. At this stage they are called *cumulus congestus*. The vertical growth is accompanied by horizontal growth as well. Then, in the afternoon, the cumulus top suddenly changes its character. Its "cauliflower" appearance becomes smooth and the top usually stretches out principally in one direction to form the anvil shape so frequently typical of the *thunderhead* or *cumulonimbus*. With the transition to the cumulonimbus comes the flash of lightning, the peals of thunder, and the surge of heavy rain.

The clouds just described are formed by the condensation of water vapor when air heated near the ground ascends and is cooled by expansion. The process of transferring heat upward by air movement is called *convection*.

**Figure 1.1** *Cumulus.* [*Courtesy of National Oceanic and Atmospheric Administration.*]

**Figure 1.2** *Cumulus congestus.* [*Courtesy of NOAA.*]

**Figure 1.3**   *Cumulonimbus. [Courtesy of NOAA.]*

Figure 1.4 shows schematically how the vertical motions of the air heated at the ground lead to the formation of cumulus clouds. The reason why rising air cools will be given in Chapter 4, and the details of the way in which cloud drops form from water vapor will be discussed in Chapter 8. The vertical extent of a convective cloud is usually about the same as its horizontal dimensions.

In contrast to these clouds of vertical development are the layer, or *stratiform*, clouds. These clouds form sheets extending horizontally hundreds and even thousands of kilometers,[1] but may be as thin as a few tenths of a kilometer and are rarely thicker than about five kilometers.

Layer clouds frequently are associated with the approach of a storm in temperate latitudes. The first sign of its approach is the advance over the sky of a sheet of very high, thin clouds, feathery *cirrus* at first, thickening to a more even layer of *cirrostratus*. The layer becomes progressively thicker and lower. When it has become sufficiently thick to obscure the outline of the sun's disc it is called *altostratus*. Gradually, the altostratus cloud becomes a uniform grey overcast through which it is no longer possible to determine the position of the sun. Finally, precipitation begins

---

[1]A kilometer (1000 meters) is 0.621 mile. For the rough order-of-magnitude considerations of the above statement the reader unfamiliar with the metric system of units can substitute approximately miles for kilometers and yards for meters. A table of equivalents of units in the metric and English systems will be found in Appendix A.

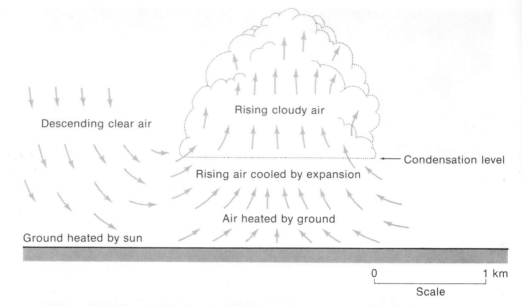

Descending clear air

Rising cloudy air

Condensation level

Rising air cooled by expansion

Air heated by ground

Ground heated by sun

0      1 km

Scale

**Figure 1.4**  *Schematic representation of the formation of a cumulus cloud by convection.*

**Figure 1.5**  *Cirrus. [Courtesy of NOAA.]*

**Figure 1.6**  *Cirrostratus with halo. [Courtesy of NOAA.]*

**Figure 1.7**  *Altostratus. [Courtesy of NOAA.]*

**Figure 1.8**  *Nimbostratus. [Courtesy of NOAA.]*

to fall, at which time the cloud is called *nimbostratus*; the rain or snow is usually continuous at this stage of the storm's development.

In winter storms over the northern United States the transition from the first appearance of cirrus and cirrostratus in the blue sky to the sullen greyness and steady precipitation of the nimbostratus usually occurs in a period of 8–24 hours. Thus if you recognize that the cirrostratus is of this type, you can forecast that rain (or snow) will begin within a day with some confidence.

This sequence of layer clouds is associated with the approach of the *warm-front* portion of a *frontal wave cyclone*. The details of this phenomenon will be discussed in Chapter 11, but a brief indication of its nature may be in order at this point.

When J. Bjerknes discovered, in 1918, that distinct boundaries occur in the atmosphere between masses of air that have different properties, and that much of the weather occurs as a result of "battles" between the air masses, he called the boundaries *fronts*. If cold air replaces warm air as a front moves, it is called a *cold front*. If warm air replaces cold air, the boundary is a *warm front*. At a warm front, however, the air doesn't merely move horizontally; the warm air moves faster than the receding cold air and climbs up along the frontal surface. The cold air doesn't get out of the way fast enough and the warm air is forced to rise over it.

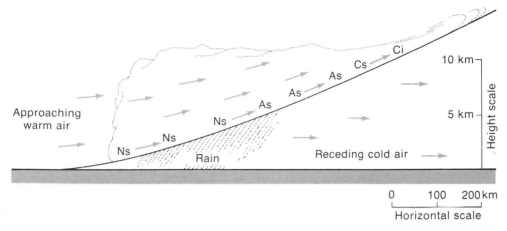

**Figure 1.9**   *Schematic representation of clouds formed by air ascending a warm front.*

Figure 1.9 shows schematically the flow of warm air over the receding cold air at a warm front and the resulting cloud distribution. As the air masses move from left to right, the front, which separates them, moves in the same direction, and the cloud pattern moves with it.[2] An observer at the right edge of the diagram would thus experience the cloud sequence that has been described herein as the warm front and its associated pattern of clouds and precipitation would move past him.

The battle of the air masses at a cold front is frequently more violent than at a warm front. The cold air pushes in under the warm air abruptly, forcing strong upward motions that are even more rapid than those occurring in a local thunderstorm. The frontal thunderstorms are more intense, and are oriented in lines along (or ahead of) the cold front. These lines of heavy thunderstorms and strong gusty winds are called *squall lines.* Tornadoes occasionally develop in the severe thunderstorms accompanying squall lines. A tornado is an intense vortex, or rotating air column, rendered visible by the funnel cloud extending downward from the base of the cumulonimbus cloud and by the swirling mass of dust and debris lifted from the ground. It has winds up to 200 m s$^{-1}$ (400 miles per hour)[3] in it and is extremely destructive. Sometimes erroneously called cyclones, tornadoes are among the most villainous characters of the meteorological drama.

The preceding discussion of clouds leads quite naturally to a classification system based on their physical characteristics, the levels at which they

----

[2]The motion of the air masses referred to is only the component of motion perpendicular to the front. There is, in addition, air flow parallel to the front, and this is usually larger. The actual wind is the resultant (sum) of these two components. Furthermore, because clouds are forming and dissipating, the movement of the cloud pattern is usually slower than the movement of the warm air relative to the front.

[3]m s$^{-1}$ ≡ meters per second. 1 m s$^{-1}$ = 2.24 miles per hour. (See table of equivalents in Appendix A.)

occur, and the processes by which they are formed. The classification now in use is an extension and modification of one introduced in 1803 by Luke Howard, an English amateur naturalist. This classification divides them into mutually exclusive genera, according to their shape and height of occurrence, species, which are subdivisions of the genera, based on additional differences in shape and internal structure, and varieties based on special characteristics within a species. In this text the identification of clouds will be limited to the genus and, in a few instances, the species.

On the basis of height, clouds are classed in four groups—high-level, middle-level, low-level, and clouds with marked vertical development—and within these groups according to whether they form layers (stratus), puffs (cumulus), or hairlike filaments (cirrus). The cirrus types are confined to high-level clouds. If there is precipitation falling from the clouds, the word nimbus is combined with stratus or cumulus: nimbostratus or cumulonimbus.

The full description of these genera and listing of species and varieties is given in the *International Cloud Atlas*, volumes I and II, published by the World Meteorological Organization. Table 1.1 outlines the basic genera and their heights and forms.

We have already discussed some of the ways in which these various clouds are formed. Further discussion will be given in connection with a more detailed study of the processes of condensation, precipitation and convection, and the large-scale atmospheric systems, in Chapters 8–11. The study of the organization of clouds, precipitation, and other weather phenomena into systems is carried out by the use of weather maps. Simplified versions of the weather map are familiar to most people from television weathercasts and their publication in newspapers. In the next section a brief introduction to this important meteorological tool will be presented.

## 1.3 Weather systems and the weather map

To determine the state of the atmosphere at any one time observations are made at weather stations throughout the world. By international agreement these observations are made simultaneously, using procedures that meet uniform standards of accuracy, and are transmitted by radio, teletype, or telegraph in a form or code that enables meteorologists of all countries to understand and use the information. Observations of the phenomena that can be evaluated by an observer at the ground without sending instruments aloft are called *surface observations*. Those obtained by instruments that are carried aloft by balloons, airplanes, or rockets are called *upper-air observations*. Nowadays, the upper-air observations that are carried out on a routine basis are made with *radiosondes* or *rawinsondes*, which are instruments that radio back the temperature, pressure, and humidity as

**Table 1.1**   Classification of Clouds

| Genus | Height (of base above ground) | Shape and appearance |
|---|---|---|
| HIGH CLOUDS | | |
| Cirrus (Ci) | 6–18 km | Delicate streaks or patches |
| Cirrostratus (Cs) | 6–18 km | Transparent thin white sheet or veil |
| Cirrocumulus (Cc) | 6–18 km | Layer of small white puffs or ripples |
| MIDDLE CLOUDS | | |
| Altocumulus (Ac) | 2–6 km | White or grey puffs or waves in patches or layers |
| Altostratus (As) | 2–6 km | Uniform white or grey sheet or layer |
| LOW CLOUDS | | |
| Stratocumulus (Sc) | 0–2 km | Patches or layer of large rolls or merged puffs |
| Stratus (St) | 0–2 km | Uniform grey layer |
| Nimbostratus (Ns) | 0–4 km | Uniform grey layer from which precipitation is falling |
| CLOUDS WITH VERTICAL DEVELOPMENT | | |
| Cumulus (Cu) | 0–3 km | Detached heaps or puffs with sharp outlines and flat bases, and slight or moderate vertical extent |
| Cumulonimbus (Cb) | 0–3 km | Large puffy clouds of great vertical extent with smooth or flattened tops, frequently anvil shaped, from which showers fall, with thunder |

relatively small, unmanned balloons carry them upward. The winds at upper levels are determined by tracking the radiosonde with radio direction-finding equipment. Because the equipment used in upper-air observations is more expensive, they are made at considerably fewer stations than surface observations.

**Figure 1.10** *Cirrocumulus. [Courtesy of NOAA.]*

**Figure 1.11** *Altocumulus. [Courtesy of NOAA.]*

**Figure 1.12**  *Stratocumulus. [Courtesy of NOAA.]*

**Figure 1.13**  *Stratus. [Courtesy of NOAA.]*

The principal elements that are observed and reported in the surface observations are the amount and kinds of clouds, the wind speed and direction, the current and past state of weather (rain, fog, etc.), the pressure, the temperature, and the humidity. The routine upper-air observations give the pressure, temperature, humidity, and wind at various heights above the station. These data are collected at regional, national, and world data centers and redistributed to various forecast centers either in their original form or in the form of *synoptic* maps and charts that summarize the weather conditions over a large portion of the earth. Until the development of high-speed electronic computers, the empirical study of synoptic weather maps formed the basis for the weather forecasts. Because of this, the branch of meteorology that is concerned with the interpretation of weather observations and weather maps is called *synoptic meteorology*.

A typical surface weather map is shown in Figure 1.14. It consists of groups of numbers and symbols at various places on the map, which represent the weather observations at those locations, and two sets of lines, the isobars (with numbers at the ends), which connect places having the same sea-level pressure, and the fronts (heavier lines with solid triangles or semicircles along them), which are the intersections with the ground of the surfaces separating different air masses. This map summarizes the weather conditions that were observed at 7:00 A.M., EST, December 15, 1971.

The complete explanation of the symbols on the weather map is given in Appendix D. Figure 1.15 is an abbreviated illustration of the interpretation of the observational data at each station. The circle, which is at the location of the station on the map, is shaded to show the fraction of the sky covered by clouds and serves as the "head" of an arrow, with a shaft extending from it in the direction from which the wind is blowing. Barbs or feathers at the other end of the shaft show the wind speed to the nearest 5 knots, a full (long) barb representing 10 knots, a half barb (short) representing 5 knots. To the upper right of the circle is a three-digit number, the pressure reduced to sea level, expressed in millibars (mb) and tenths, with the first one or two digits omitted. Since the pressure at sea level is always on one side or the other of 1000 mb and varies gradually for the most part, you can readily tell what the missing digit(s) is.

To the left of the station circle at the same level as the pressure is the temperature, given in degrees Fahrenheit on surface maps in the United States. This procedure is a deviation from international practice; on the upper-air charts the temperatures are given in degrees Celsius (centigrade) in the United States as they are elsewhere in the world. To the left, a little lower than the station circle, is the dew-point temperature, also in degrees Fahrenheit, and immediately to the left of the circle, between the two temperatures, is a symbol describing the character of the weather at the time of observation, groups of dots if it is raining, stars if it is snowing, and so forth. (See the chart in Appendix D for the 100 different symbols that can be used.)

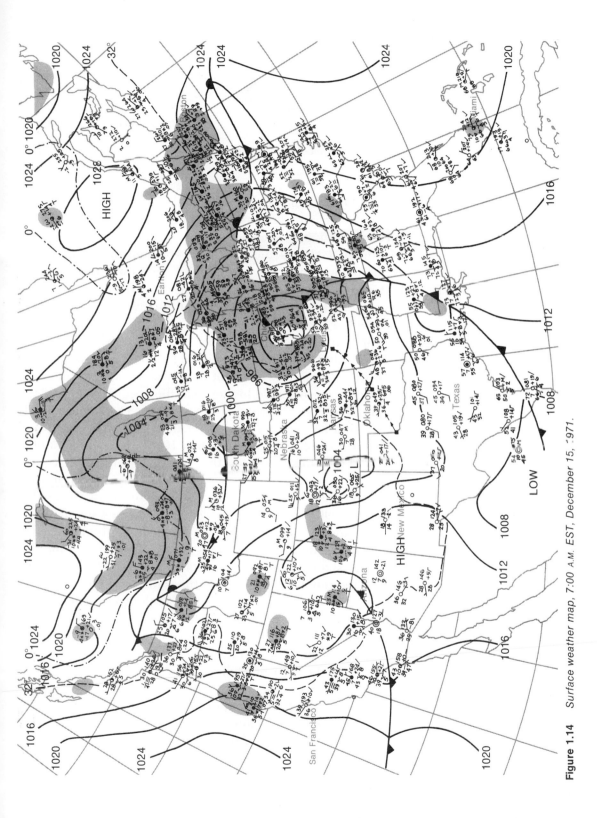

**Figure 1.14** *Surface weather map, 7:00 A.M. EST, December 15, 1971.*

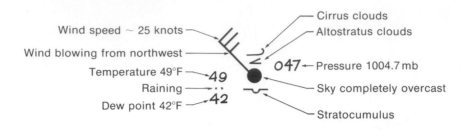

**Figure 1.15** *Example of interpretation of symbols on weather map.*

Above and below the station circle are symbols that show what kinds of high, middle, and low clouds are present. Other numbers and symbols, not shown in Figure 1.15, give such additional information as the trend of pressure during the past 3 hours, the state of weather and amount of precipitation during the 6 hours previous to the observation time, the visibility, and the height of the base and amount of the lowest cloud. Thus the weather map enables us to see what the conditions were at the various stations in considerable detail.

The fact that pressure varies gradually and continuously from place to place permits its distribution to be summarized by drawing isobars—lines of constant pressure that are drawn so that the pressure on one side of the line is lower and on the other side higher than the specified value. Thus in Figure 1.14 the line going southward across South Dakota and southeastward across Nebraska and northeastern Kansas has pressures lower than 1000 mb to the east of it and higher than 1000 mb to the west of it. Following this 1000-mb isobar, we see that it forms a closed curve, enclosing a large area of lower pressures. The isobars are drawn at 4-mb intervals, and within the 1000-mb isobar there are successively smaller areas enclosed by the 996-, 992-, 988-, and 984-mb isobars. A low-pressure area of this sort is called a *cyclone*. The winds, as shown by the arrows, while tending to some extent to blow across the isobars toward the low center, by and large show a tendency to circle counterclockwise around the center—east of the center they flow from the south, north of it from the east, west of it from the north, and south of it from the west. The stations near the low center had overcast skies, and it was raining at most of them at the time of observation. The winds were relatively strong and the weather unpleasant near the center of this cyclonic storm.

In contrast, consider the area of high pressure over Arizona and New Mexico. There the skies were clear and the winds light. High-pressure areas are frequently regions of fair weather. On this map the high centers are not well enough defined to show the typical pattern of wind flow around them, which is clockwise (in the Northern Hemisphere). Having characteristics opposite to those of the cyclones, the high-pressure areas and associated wind systems are called *anticyclones*.

Cyclones and anticyclones move and their intensity changes with time, but, in general, they maintain their identity from day to day. Their movement and change is associated with the change in weather at any particular location. The problem of weather forecasting may be considered to be the prediction of the movement and development of these pressure systems and their associated weather patterns. The movement of the low center for the 24 hours previous to the map time for Figure 1.14 is shown by the string of arrows extending southwestward from it to the Texas-Oklahoma border. As a first guess we might assume that the movement in the next 24 hours would be the same distance in the same direction. This would place the low just east of Earlton, Ontario at 7:00 A.M., EST, December 16. We would then guess that as it moved the area in which it had been would experience improving weather, with the rain stopping, the skies tending to clear, and the winds becoming lighter and shifting to northwest (corresponding to the low being to the northeast of its previous position).

The surface weather map for 7:00 A.M., EST, December 16 is shown in Figure 1.16. While the low center has moved somewhat faster than it had during the previous 24 hours, its position is not very far from our estimate. As we guessed, the rain has stopped and the winds have shifted to northwest in the region where the low passed, but contrary to expectations, a number of the stations there continue to have low clouds covering the sky. Thus our first attempt at forecasting, using the weather maps, has had some success, but also, as happens even when more sophisticated methods are used, some degree of failure.

## 1.4   New tools for meteorology

The weather map remains one of the key tools for the study of atmospheric processes and the prediction of the weather. Recently, two new devices, familiar to everyone through news media, which serve to increase our knowledge of the state of the atmosphere and our ability to carry out the forecasting process, have been introduced. They are the meteorological satellites and the high-speed electronic computers.

With the launching on April 1, 1960, of TIROS I, the first artificial satellite transmitting television pictures of cloud patterns back to the earth, a new era in meteorology began. Previously, the patterns of clouds and storms were inferred only from the separate stations taking surface observations. The clouds seen by the observers at these stations are only those within a few tens of kilometers of the stations. Since the stations are 200 km or more apart over continents, even over these areas the synoptic weather map gives only a partial picture of the cloud patterns.

The satellite pictures give an overview of areas about 1000 km across or more at a time, and these pictures have shown patterns that had not been recognized or had been appreciated only partially previously. Some

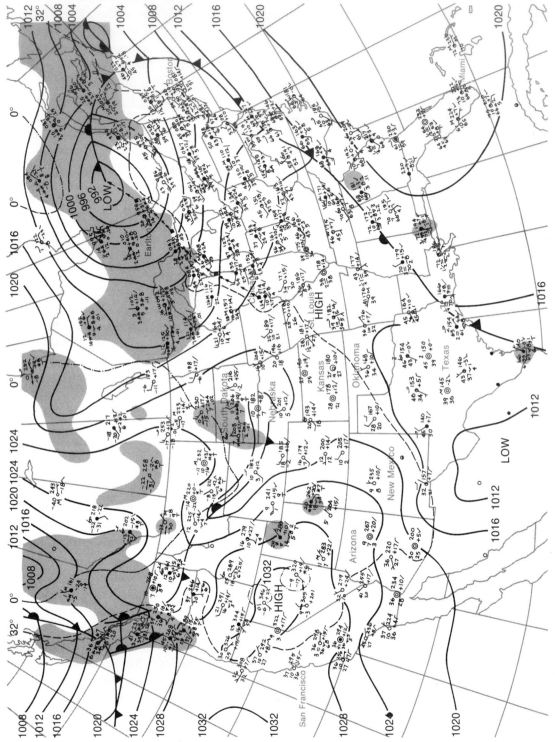

**Figure 1.16** *Surface weather map, 7:00 A.M. EST, December 16, 1971.*

examples of these are the banded structure of the clouds around large, middle-latitude cyclones, the occurrence of eddies in the lee of islands, and the occurrence at times of large "hollow" cells of convective clouds over oceans.

Even more important than the fact that the satellites give an overall view and show the distribution of clouds in the areas between stations in places where there are observing posts is the fact that they provide information for the large areas of the earth in which no stations exist. For instance, only a few island stations and weather ships, together with ocean-going vessels and airplanes in flight along established trade routes, supply the data we have for weather conditions over the seven-tenths of the earth covered by oceans. Large portions of the earth are completely devoid of observations from within the atmosphere. For these areas the satellites provide the only information presently available. On some occasions they have given evidence of the existence and location of hurricanes and typhoons long before they would have been located otherwise. Particularly for such places as Australia, where the weather systems come from the South Indian Ocean and the Southern Ocean, where few ships travel, the cloud pictures from satellites have been especially helpful in making the daily weather forecasts.

The early TIROS satellites were spin-stabilized and launched in orbits inclined about 50° to the equator. These orbits gave only partial coverage of the earth, and the spin-stabilization with the cameras pointing along the axis of spin meant that part of the time they were looking at the sky rather than the earth. On the average, the TIROS took pictures of about 20 percent of the earth each day. In 1966 a new series of satellites, called ESSA, were launched on an operational basis, in a polar orbit with earth-orientation. These satellites cover most of the earth during the daylight hours. Furthermore, experimental satellites in the Nimbus series are equipped with infrared radiation devices that enable the determination of cloud cover at night as well as in the daytime.

Beginning in December 1966, synchronous satellites have been operational in stationary positions above the equator, being shifted from one longitude to another on instruction by radio from the earth. When a satellite is in orbit in the plane of the equator at a height of about 23,000 miles it moves around the earth at the same rate at which the earth turns and thus remains over the same place on the earth's surface. Thus ATS I stayed over a spot in the middle of the Pacific Ocean taking pictures for several years. ATS III, which was launched in November, 1967, was first maintained over the mouth of the Amazon River, and subsequently moved back and forth between there and the mid-Pacific, spending periods of several months in each position. The picture in the frontispiece was taken from ATS III when it was over western South America, near the headwaters of the Amazon River. In pictures like this, an area of 225,000,000 km², almost one-half of the earth's surface, is seen. Photographs taken at frequent intervals enable changes in cloud formations over large areas to be watched practically continuously, whereas previously no information at all was available from many of these areas.

While the cloud photographs from satellites fill in the gaps in coverage of the meteorological scene, they have limited value in forecasting the development of the drama of the weather. As we shall see, the procedures used in weather forecasting require knowledge of the wind, temperature, humidity, and pressure at various heights over large areas, in addition to the distribution of clouds and precipitation. The cloud formations and radiation measurements from satellites give some information about the other needed quantities, but so far only to a limited extent. From the positions of clouds on successive pictures from geostationary satellites an approximation of wind speeds at cloud levels can be estimated. Satellite measurements of infrared radiation from the atmosphere obtained from recent experimental Nimbus series satellites have enabled estimation of the temperature variation in the upper layers of the atmosphere. These techniques are just beginning to be used in routine operations. As methods are improved to provide accurate satellite measurements of wind velocity, temperature, pressure, and humidity within the atmosphere, they will be employed directly in dynamic weather prediction using high-speed computers.

Another approach to obtaining the worldwide distribution of temperatures and winds at upper levels is the proposal to track by satellite the movement and temperature measurements of large numbers of long-lasting balloons that float at constant elevations above the earth. In preliminary tests such balloons have been launched from New Zealand and tracked by ground-based radio-direction finding equipment. The tests (called GHOST, for Global Horizontal Sounding Tests) were conducted in the Southern Hemisphere to reduce the need to obtain permission for the balloons to traverse other countries, and to minimize the possibilities of collision with aircraft, even though the balloons are so small that they do not constitute a danger to airplanes. The balloons designed to float at about 12 km (40,000 ft) were very successful, some remaining aloft for many months and circling the earth several times. At the lower levels that were tried, the balloons apparently tended to pick up water and ice from clouds and to fall into the ocean.

The use of high-speed electronic computers is the other very important advance in meteorology during the past two decades. In fact, the whole process of weather prediction has been completely revolutionized by its use. Previously, the entire procedure was empirical and highly subjective. The reason for this is that the equations governing the behavior of the atmosphere are so extremely complex that they cannot be solved by ordinary mathematical procedures. With the high-speed computers, methods have been devised for finding approximate solutions, at least so far as the larger motions are concerned. This has led to a situation in which a large part of the prediction can be carried out by machine with a fair degree of accuracy. There remain parts of the forecasting, however, that are still subject to human error, in addition to the uncertainties due to the gaps in the observed state from which we start and the simplifications required to obtain solutions, even with the computer. But efforts are increasing to improve the observational network and the computational methods. We can expect these efforts to be reflected in improvements in the accuracy of the weather forecasts in the not-too-distant future.

## 1.5　The role of mathematics in science

The reference to equations in the preceding section brings up a subject that is a psychological block for many nonscience students. Having somehow surmounted the required mathematics courses in high school, they have put out of mind what they learned of it there and have made no attempt to use it in their daily lives. But much of our daily experience consists of quantitative relationships that are by their nature mathematical, and this is especially true of the scientific aspects of everyday life. While many of these relationships can be expressed in words, they are more easily and concisely expressed in numbers, symbols, and equations. In this sense, mathematics is the *language* of science. In addition to clarity and ease of expression, this language has the virtue of being universal. Scientists of practically all nationalities, however different their native tongues, use and understand it. Whatever the language and alphabet used in the rest of their writing, American, French, Russian, Chinese, or Japanese mathematicians, physicists, and meteorologists use the same systems of equations and symbols in the mathematical part.

In addition to serving as the language of science, mathematics provides a method of reasoning that enables the discovery of new relationships. Thus in physics, mathematical rules are set up that correspond to processes occurring in the real physical world, and these rules enable deductions to be made concerning the consequences of these processes under prescribed conditions. We shall see how this is done in our analysis of some atmospheric processes.

In this book the quantitative relationships will be presented both in words and in mathematical symbols. The mathematics used will be that usually required for admission to college, namely, high school algebra or the equivalent. Both from the standpoint of appreciating the methods of atmospheric scientists in applying the laws of physics to atmospheric phenomena and from the standpoint of being better able to understand and remember the atmospheric relationships that will be discussed in the following chapters, it is desirable that the reader study the mathematical treatments presented. However, mastery of the equations is not necessary; the concepts can be obtained from the verbal presentation.

## Questions, Problems, and Projects for Chapter 1

1   Write down for yourself, as frankly as you can, the reason why you are studying meteorology, and list the questions you expect to learn the answers to, or the topics you expect to find out about.

2   Keep a log of the cloud types you see every day. If convenient, choose two or three times of the day when you are out of doors regularly (e.g., on your way to class), and note the kinds of clouds in the sky at those times and your estimate of the fraction of the sky covered by each.

3   List the phenomena that you consider constitute *weather*. From these formulate a definition of weather.

4   Express the cloud heights given in Table 1.1 in feet.

5   From the data in Figure 1.14 write a description of the weather at 7:00 A.M., December 15, 1971, at Boston, Massachusetts; Chicago, Illinois; St. Louis, Missouri; Miami, Florida; and San Francisco, California.

6   What is meant by "fronts" in meteorology? Why do cloudiness and precipitation frequently occur in their vicinity?

7   What innovations have been introduced by modern technology in the study and prediction of the weather? What benefits have resulted from these innovations?

# TWO

# The Composition and Thermal Structure of the Atmosphere

## 2.1 Nature and composition of the atmosphere

The atmosphere of the earth is a gaseous envelope that surrounds the solid and liquid surface of the earth. It extends upward for hundreds of kilometers, eventually meeting with the rarefied interplanetary medium of the solar system, which in turn may be considered an extension of the corona of the sun.

The phenomena that take place in the atmosphere are many and varied. In the lower layers winds blow, clouds form and dissipate, rain and snow fall, and warm and cold spells occur. These are among the phenomena that are commonly called *weather*. At higher levels the aurora borealis, zodiacal light, luminous meteors, and noctilucent clouds are seen, and radio signals are reflected or refracted by the ionosphere. Particles in the atmosphere cause visible effects, such as rainbows, halos, colors of the sky, and the changes in the appearance of the landscape due to haze. Invisible but equally important effects are caused by chemical (or photochemical) and electrical reactions between atmospheric constituents. The study of all these phenomena is the subject of meteorology.

The gas that constitutes the atmosphere is called *air*. Air is a mixture of several chemical elements and compounds. Some early Greek philosophers thought that air was the primary element, that is, the fundamental substance not further subdivisible into constituent components from which everything derived. Empedocles thought that the universe was composed of four such elements: air, fire, earth (soil), and water. We now know that there are many elements (103 have been discovered so far) and, further, that in addition to elements the air contains some compounds, that is, chemical combinations of elements. The most plentiful elements in air are nitrogen, oxygen, and argon. The compounds in it of greatest importance are water vapor and carbon dioxide. Another important element, even though it is present only in small amounts and mostly at high elevations, is ozone, the triatomic form of oxygen.

Water vapor is constantly being added to the atmosphere by evaporation from various bodies of water or damp ground and by transpiration from plants; it is being subtracted from it in other places by condensation into clouds or dew. For this reason the amount of water vapor present is quite variable. Over warm oceans or tropical jungles it may be as much as 3 or (rarely) 4 percent of the air by volume; in cold regions, over deserts, or at great heights, it is a small fraction of 1 percent.[1]

Because of this variability, it is usual to consider separately the water-vapor content and the fractional composition of the remainder of the air (which for convenience we call *dry air*). It has been found that the composition of dry air is almost exactly the same all over the earth and to heights up to about 100 km. Table 2.1 gives the percentages of the major components, nitrogen, oxygen, argon, and carbon dioxide. The other constituents, which include the inert elements neon, helium, krypton, and xenon, as well as hydrogen and some compounds, such as methane, sulfur dioxide, and various oxides of nitrogen, that come from biological and industrial processes, are present in extremely small fractions, a few parts per million or less.

In addition to the gaseous components, air contains solid and liquid particles, most of them so small that the movements of the air offset their tendency to fall to the ground. These may originate as dust raised by the wind, as sea salt particles from the evaporation of ocean spray, or as products of combustion and industrial processing. There are also particles that come into the atmosphere from extraterrestrial sources.

Our knowledge of the composition of air dates from the beginning of modern chemistry in the eighteenth century.[2] Prior to that time, John Mayow (1643–1679) conducted experiments that led him to conclude that air consists of two parts: one, called "fire-air," that supports combustion and sustains life and another that does not. However, he did not isolate the separate constituents.

---

[1] It is shown in kinetic theory that for a given pressure and temperature the same number of molecules of any gas, whatever the mass or size of its molecules, occupies the same volume. When we speak of "percent by volume" we are thus identifying what percent of the total number of molecules are molecules of the particular constituent.

[2] The account given here is condensed from that published in *Ways of the Weather*, by W. J. Humphreys (Lancaster, Pa.: Jaques Cattel Press, 1942).

**Table 2.1**    Composition of Dry Air

| Constitutent | Chemical symbol | Content (% by volume) |
| --- | --- | --- |
| Nitrogen | $N_2$ | 78.08 |
| Oxygen | $O_2$ | 20.95 |
| Argon | A | 0.93 |
| Carbon dioxide | $CO_2$ | 0.03 |
| Total | | 99.99+ |

Mayow's discovery was overlooked, and, in fact, the first constituent of air to be isolated and studied in detail was not "fire air" (oxygen) or the more plentiful nitrogen, but carbon dioxide. This work was carried out by Joseph Black (1728–1799) in 1752, when he was a medical student at the University of Edinburgh. Black called the substance that he isolated "fixed air," because he found it to be somehow fastened to various substances (carbonates) from which it could be removed by heating.

The next advance, the discovery of nitrogen, was also made at Edinburgh by a medical student named Daniel Rutherford (1749–1819). Working under Joseph Black in 1772, he studied the properties of the gas that remained after charcoal was burned in a closed volume for as long as possible and then the carbon dioxide was removed. He called this gas "mephitic air" and thought that it was a combination of ordinary air and "phlogiston," the material considered to be given off by substances in the process of burning.

Shortly after Rutherford isolated "mephitic air," Joseph Priestley (1733–1804), in England, and Carl William Scheele (1742–1786), in Germany, isolated oxygen. Priestley did so by heating mercury in the presence of air until it formed a red powder and then increasing the heat until a gas was given off that supported combustion more strongly than ordinary air. He called this gas "dephlogistated air."

Subsequently, the nature of these components of air was clarified, and they were given their modern names. But it was not until 1894 that argon, which is almost 1 percent of dry air, was identified. Until that time it was lumped with nitrogen, but in that year Lord Rayleigh (1842–1919) recognized that "nitrogen" isolated from air had a heavier molecular weight than nitrogen obtained from the decomposition of chemical compounds. He and Sir William Ramsay (1852–1916) carried out experiments in which the nitrogen was removed from air. They found an inert residual gas, which they named argon.

The discovery that air contained a hitherto unknown constituent in such a large proportion stimulated more careful analysis to see whether other constituents were present. It was found that four additional inert gases—helium, neon, krypton, and xenon—are present in very small quantities, and that hydrogen, which is given off at the earth's surface and escapes upward to space, is present in a similarly small concentration.

In addition to these "permanent" gaseous constituents of the atmosphere there are many variable constituents introduced by biological activities and industrial processes. Among the principal variable constituents are sulfur dioxide ($SO_2$), oxides of nitrogen ($N_2O$, $NO$, $NO_2$), ammonia ($NH_3$), methane ($CH_4$), carbon monoxide ($CO$), ozone ($O_3$), and various organic compounds. Some of these substances are important air pollutants, having deleterious effects even when present in concentrations as low as one part per million or less.

Accurate measurements show that carbon dioxide also is a variable constituent. Not only is its concentration higher near urban and industrial complexes, but its average volume percentage has increased steadily from about 0.029 percent in 1900 to 0.033 percent today. It has been suggested that this increase has influenced the way in which the average temperature of the earth has changed since the beginning of the century, and that further increase in its concentration may have catastrophic consequences by causing the thick ice sheets that cover Greenland and Antarctica to melt, thereby raising the sea level enough to inundate coastal cities throughout the world. Carbon dioxide tends to raise the earth's temperature because it absorbs and radiates back some of the infrared radiation emitted by the ground. This additional heating makes it necessary for the ground to be warmer in order to emit enough radiation to balance the radiation received from the sun. Counteracting this effect of carbon dioxide, particles from volcanoes and pollution sources may significantly reduce the amount of solar radiation reaching the ground, thereby tending to decrease the earth's temperature.

## 2.2 Vertical structure of the atmosphere

From common experience we know that the temperature of the air varies from time to time: warm afternoons follow cool mornings and warm summers follow cold winters in a regular cyclical fashion, and spells of cold days and warm days are interspersed in a single season in an irregular fashion. The temperature also varies from place to place at the same time; in general, regions at low latitudes are warmer than those at higher latitudes, and low lands are warmer than high mountains. The familiar Japanese wood prints showing trees in blossom with snow-capped Mt. Fuji towering in the background illustrate the effect of altitude on temperature.

If we consider the earth as a whole for the entire year, the average temperature near the ground (the surface temperature) is 15°C (288°K, 59°F).[3] This year-round overall average decreases with height, as indicated

---

[3]In this book we shall usually express temperatures on the Celsius (centigrade) scale or the Kelvin (absolute centigrade) scale rather than the Fahrenheit scale, which is commonly used in the U.S. Rules for converting from one of these scales to any other are given in Appendix A, and a discussion of them and the relationships for conversions is given in Chapter 4.

in the previous paragraph, but above about 12 km (40,000 ft) it stops decreasing and starts increasing. This surprising reversal was not believed when it was first discovered by sending up small balloons with recording instruments. At first it was thought to be an instrumental error, and only after several hundred balloons had reached these heights and the same result was obtained was the existence of this layer of inverse behavior of temperature accepted.

The bottom layer of the atmosphere, in which the temperature decreases with height, is called the *troposphere*; the layer above, of constant or increasing temperature, is called the *stratosphere*; and the surface between them (sometimes a transition zone or layer rather than a surface) is the *tropopause*. In the troposphere the temperature on the average decreases at a fairly constant rate, 6.5°C/km (3.5°F/1000 ft).

Early in the twentieth century various pieces of indirect evidence, including the way in which some large explosions were heard both close to the places where the explosions occurred and at greater distances from them, with zones of silence between, led to the deduction that at a height of about 50 km the temperature is about the same as at the earth's surface; other indirect evidence pointed to very low temperatures at about 80 km, and still other evidence indicated that above 80 km the temperature increases again. Thus, before the direct observations by rockets and satellites the general atmospheric structure was known to considerable heights.

The use of rockets and satellites for the exploration of the high atmosphere has confirmed this knowledge and provided precise measurements of the temperature to much greater heights. After the confirmation of the existence of the temperature maximum at about 50 km and the minimum at about 80 km, names were given to additional layers. The layer of decreasing temperature from 50 to 80 km is called the *mesosphere*, and the layer of increasing temperature above that is called the *thermosphere*. The boundary between the stratosphere and mesosphere is called the *stratopause*; that between the mesosphere and thermosphere is called the *mesopause*.

Other designations of atmospheric layers are based on electrical state and on composition. The *ionosphere* is a layer that contains ions (molecules and atoms that carry electric charges) and free electrons. It extends upward from about 80 km and measurements of radio reflections indicate that it is composed of layers of varying ion density, the E- and F-layers. From the standpoint of composition, sometimes the layer of uniform composition with respect to major constituents, which extends from the earth's surface to 80 or 100 km, is called the *homosphere*; the layer above that, where chemical species are different at various heights, is called the *heterosphere*. Between 15 and 50 km the concentration of ozone, although very small compared with that of oxygen and nitrogen, is relatively large. This ozone-rich layer is called the *ozonosphere*.

Particles as high as 60,000 km, acted on by the earth's magnetic and gravitational fields, mostly move with the earth and can therefore be con-

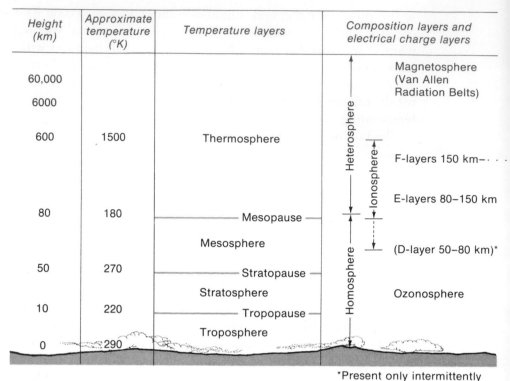

| Height (km) | Approximate temperature (°K) | Temperature layers | Composition layers and electrical charge layers |
|---|---|---|---|
| 60,000 | | | Magnetosphere (Van Allen Radiation Belts) |
| 6000 | | | |
| 600 | 1500 | Thermosphere | F-layers 150 km–··· |
| | | | E-layers 80–150 km |
| 80 | 180 | Mesopause | |
| | | Mesosphere | (D-layer 50–80 km)* |
| 50 | 270 | Stratopause | Ozonosphere |
| | | Stratosphere | |
| 10 | 220 | Tropopause | |
| | | Troposphere | |
| 0 | 290 | | |

Heterosphere · Ionosphere · Homosphere

*Present only intermittently

**Figure 2.1** *Layers of the atmosphere.*

sidered to be part of the earth's atmosphere. In this outermost layer are the *Van Allen belts* of charged particles. This layer, in which the movement of the charged particles is largely controlled by the magnetic field, is called the *magnetosphere*.

The vertical structure of the atmosphere, based on temperature, composition, and electrical state, is shown in Figure 2.1. The heights and temperatures given in Figure 2.1 are rough averages characteristic of middle latitudes where most of the observations of temperatures aloft have been made. As an indication of the degree of approximation the heights are given only to one significant figure,[4] and the temperature to two. To this degree of approximation the data in the figure may be considered to represent the annual averages over the entire earth.

Approximate average temperature conditions at various heights over the earth are given in the *U.S. Standard Atmosphere.* This is an idealized

---

[4]By "significant figures" is meant the digits (other than the zeros adjacent to the decimal point on either side) in a number that are known accurately. Thus the statement that the height of 50 km in the figure is given only to one significant figure means that we know that it is closer to 50 km than to 40 km or 60 km, but that the accurate value might be anything between 45 and 55 km.

representation of the conditions that are considered to be typical for middle latitudes, say 45° north. It was arrived at because of the need to define a reference atmosphere for use in the design of aircraft and missiles and the instruments used on them. Up to 20 km it incorporates the *International Standard Atmosphere* agreed on by the International Civil Aviation Organization (ICAO). For greater heights where there are fewer observations, a committee of American scientists arrived at estimates of the appropriate values. When more data become available these values will doubtless be revised.

The temperatures of the U.S. Standard Atmosphere are given in Table 2.2 for heights to 230 km, and in Figure 2.2 they are shown graphically to 120 km. In the standard atmosphere the temperature in the troposphere decreases at a constant rate of 6.5°C/km to 11 km. The stratosphere consists of three sub-layers: isothermal from 11 to 20 km, temperature increasing from 20 to 47 km, and isothermal again from 47 to 52 km. The mesosphere, with temperature decreasing again, extends from 52 to 79 km, and the thermosphere begins with an isothermal layer, above which the temperature rises with height at an increasing rate. Thus the tropopause in the standard atmosphere is at 11 km, the stratopause is at 52 km, and the mesopause is at 79 km.

The accumulation of our present knowledge of the thermal structure of the upper atmosphere was a gradual process. Apart from measurements at a few mountain-top observatories, it began with the raising of thermometers on kites in the middle of the eighteenth century and the invention of the lighter-than-air-balloon at the end of that century. J. L. Gay-Lussac (1778–1850), the French scientist who later gained fame for his research in physics and chemistry, made two balloon ascents in 1804, the second of which reached 7 km. He measured the temperature and humidity up to this height and collected samples of air at different heights. In his analysis of the air samples he found no detectable variation with height.

**Table 2.2**    Temperatures of the Standard Atmosphere

| Height (km) | Temperature (°K) | Height (km) | Temperature (°K) |
|---|---|---|---|
| 0.000 | 288.15 | 90 | 180.65 |
| 11.000 | 216.65 | 100 | 210.02 |
| 20.000 | 216.65 | 110 | 257.00 |
| 32.000 | 228.65 | 120 | 359.49 |
| 47.000 | 270.65 | 150 | 892.79 |
| 52.000 | 270.65 | 160 | 1,022.2 |
| 61.000 | 252.65 | 170 | 1,103.4 |
| 79.000 | 180.65 | 190 | 1,205.4 |
| 88.743 | 180.65 | 230 | 1,322.3 |

Source: *U.S. Standard Atmosphere* (Washington, D.C.: U.S. Government Printing Office, December, 1962).

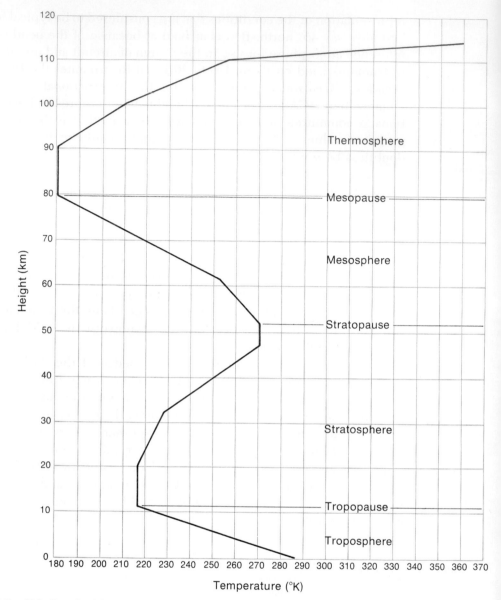

**Figure 2.2** *U.S. Standard Atmosphere.*

There were further measurements by manned balloon, particularly during the latter half of the nineteenth century (in 1862 Glaisher reached an estimated height of 11.2 km), but systematic studies began toward the end of the century with the work of R. Assmann (1845–1918) and L. P. Teisserenc de Bort (1855–1913). Assmann invented the aspirated psychrometer, which consists of wet-bulb and dry-bulb thermometers over which the air is drawn by a built-in fan to eliminate the radiational errors to which most previous temperature and humidity measurements on manned balloons were subject.

**Figure 2.3**  *Assmann ventilated psychrometer. A spring-driven motor, wound by the key at the bottom, operates a fan that draws air across the bulbs of two thermometers. The bulb of one of the thermometers is covered with a muslin wick, which is moistened with distilled water. The wet-bulb thermometer is cooled by evaporation below the temperature shown by the dry-bulb thermometer. The humidity is evaluated by computation or taken from a table, using the readings of the two thermometers and the pressure. [Courtesy of Science Associates, Inc., Princeton, New Jersey.]*

The next important advance was made when Assmann and Teisserenc de Bort developed meteorographs that could be sent aloft on unmanned balloons. In some meteorographs the pressure, temperature, and humidity were automatically recorded on a drum rotated by clockwork. In others the temperature and humidity were recorded on a slide moved by a pressure-sensing unit. The balloons burst at great heights and the meteorographs descended by parachute to the ground. When the instruments were recovered, the records were evaluated. Using unmanned balloons, Teisserenc de Bort discovered in 1898 and Assmann subsequently confirmed that the temperature stopped decreasing upward above a height ranging from 8 to 12 km in different situations. The region above this, now called the *stratosphere*, was termed by them the isothermal region.

In the early part of the twentieth century, upper-air observations were begun on a routine basis to obtain data for use in weather forecasting. For this purpose recording meteorographs were flown on huge box kites, captive balloons, and airplanes. This enabled the immediate evaluation of the record at the end of the flight, eliminating the wait until the parachuted records from the unmanned balloon flights were found. However, the heights reached were limited to the lowest five kilometers.

During the period 1928–1937 the radio-meteorograph[5] was introduced. With this device the temperature, pressure, and humidity at each level is immediately transmitted by radio to the observer at the ground so that the sounding from an unmanned balloon can be evaluated while it is ascending, making the recovery of the meteorograph unnecessary. Since its invention, systematic daily soundings have been made at an increasing number of points, and the balloons and instruments have been improved so that heights above 30 km are reached regularly. As these observations accumulated, it became clear that the apparent isothermal region always merges into a region of increasing temperature, and at low latitudes, where the

---

[5] Now called the *radiosonde*.

troposphere extends to 16 or 17 km, there is no isothermal layer, the temperature rise beginning immediately at the tropopause. At first, the term stratosphere was limited by some to the isothermal layer, and the layer in which the temperature increases with height and then decreases to the temperature minimum at 80 km was called the mesosphere. The view of those who advocated calling the entire layer through which the temperature does not decrease the stratosphere (including the isothermal regions and the region of increasing temperature) has prevailed, but in some articles, books, and reference works (dictionaries and encyclopedias) you may still find the older definition.

As stated earlier, the existence of high temperatures at about 50 km, a temperature minimum at about 80 km, and higher temperatures in the ionosphere had already been determined on the basis of indirect measurements before the means for carrying instruments to these great heights became available. This took place principally in the 1920's and 1930's. The use of rockets after World War II and artificial satellites since 1958 has made possible the evaluation of the properties of air at great heights from data obtained at those levels.

## 2.3  Latitudinal and seasonal variations of temperature at various heights

The general characteristics of temperature near the earth's surface are familiar to everyone: cold in polar regions, particularly in winter, seasonal changes in middle latitudes, from hot summers to cold winters, and warm in equatorial regions throughout the year. Is the variation with latitude and season the same at higher levels in the atmosphere? To examine this question the variation of temperature with height at different latitudes and seasons are graphed in Figures 2.4, 2.5, and 2.6.

The typical conditions in January and July at various latitudes in the Northern Hemisphere are given in the U.S. Standard Atmosphere Supplements. Curves for the data for January at latitudes of 30°N, 45°N, and 60°N are shown in Figure 2.4, together with the annual average curve for 15°N, where there is not much change through the year. Up to about 10 km it is warm at low latitudes and colder towards the pole, the same as at the ground. While the tropopause at high latitudes is low (8 km at 60°N), at 15°N and 30°N the temperature continues to decrease up to 16 or 17 km. Consequently, the temperature at levels between about 10 and 22 km is lower at low latitudes than at high latitudes. Above the level of the tropical tropopause the temperature increases with height at a greater rate in low latitudes than nearer the poles and the difference between the temperature at low latitudes and that at high latitudes decreases again. In the upper part of the stratosphere it reverses again so that the pattern in winter is similar to the lower troposphere.

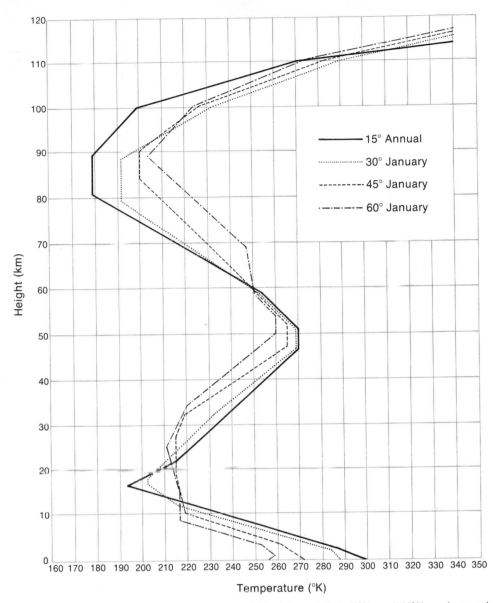

**Figure 2.4** *January average variation of temperature with height at 30°N, 45°N, and 60°N, and annual average at 15°N.*

Figure 2.5 shows the curves for the same latitudes for July. The tropopause at 60°N is a little higher (10 km) in July than in January and there is little difference between 15°N and 30°N throughout the troposphere, but the main effect, the reversal of temperature variation above the height of the polar tropopause, is present in July as well as in January. In July the temperature is higher at high latitudes than at low latitudes throughout the stratosphere.

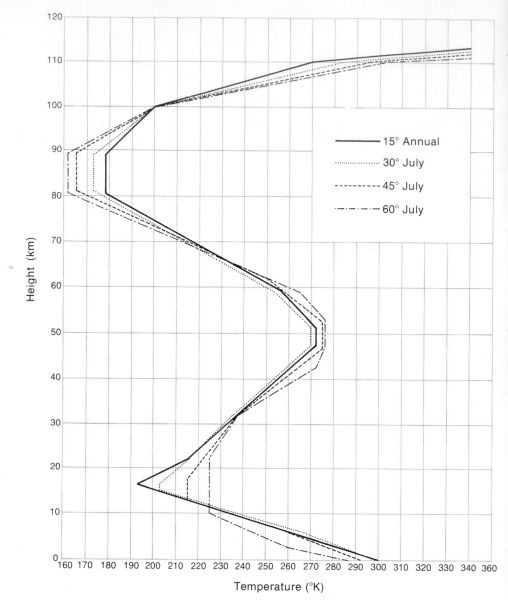

**Figure 2.5** *July average variation of temperature with height at 30° N, 45° N, and 60° N, and annual average at 15° N.*

Figure 2.6 is a different kind of representation. In this diagram the horizontal coordinate is latitude, the vertical coordinate height, and the temperature distribution is shown by isotherms — continuous lines drawn at 10-degree intervals of temperature, separating higher temperatures than the given value on one side of the line from lower temperatures on the other. Thus the diagram shows the variation of temperature with height at all latitudes, separately for winter and summer averages around the earth. The tropopause and stratopause are shown by dashed lines. In this

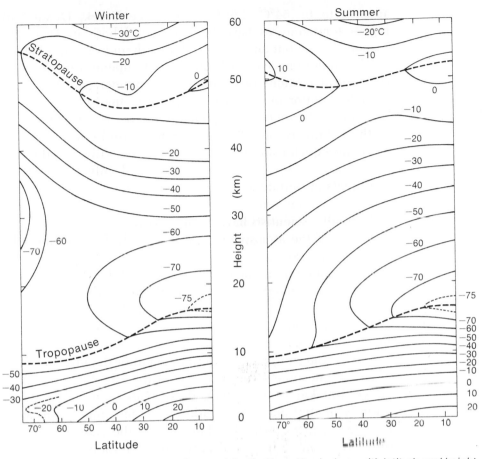

**Figure 2.6**  *Variation of temperature averaged around the Northern Hemisphere with latitude and height, winter and summer. [Lower part of figure based on data from H. Wexler, "Annual and Diurnal Temperature Variation in the Upper Atmosphere," Tellus 2 (1950): 263; upper part based on data from A. Kochanski, "Circulation and Temperature at 10- to 100-Kilometer Height," J. Geophys. Res. 68 (1963): 222.]*

figure we see that the tropopause slopes upward from arctic to tropical regions both in winter and in summer, with the isotherms through most of the troposphere having a similar slope, corresponding to lower temperatures in arctic regions. The reversal of the horizontal temperature gradient at heights that are in the troposphere in low latitudes and in the stratosphere at high latitudes is readily seen. The temperature at the tropopause near the equator is the lowest shown on the diagram.

The only place at the earth's surface where temperatures as low as the equatorial tropopause are observed is on the Antarctic continent, near the South Pole, in the southern winter. The lowest temperature ever observed in the air at the earth's surface was −87°C (−125°F) at Vostok, Antarctica. The highest temperatures on earth occur, not at the equator, but in the deserts of subtropical latitudes in summer. The record high temperature appears to be 54°C (129°F) at Furnace Creek, Death Valley. Slightly higher temperatures have been reported both in Death Valley and in Tripoli, Libya, but analysis has shown them to be unreliable.

The long-term average surface air temperatures are also higher over subtropical portions of continents in summer, and lower over the high-latitude interior of continents in winter, than elsewhere. Figure 2.7, in which the distribution of average temperatures near the ground over the earth in January and July is shown by isotherms, shows these effects. Near the equator and over the oceans generally, the difference between the average temperatures in January and July is much less than over continents at middle and high latitudes. The moderating influence of the ocean affects the temperatures of the adjoining coastal areas.

Naturally, scientists have been concerned not only with determining the structure of the atmosphere, but also with the question why? Why does the temperature vary as it does in the vertical and horizontal? Why is the mean temperature at the earth's surface 288°K? Why does the temperature decrease with height in the troposphere, and why at the particular rate that is observed? Why does it increase in the stratosphere, decrease in the mesosphere, and increase in the thermosphere? Why is the temperature higher in summer at subtropical latitudes than at the equator? Why do the extreme values occur over continents rather than the oceans?

Other questions concern the ionosphere. Why does the atmosphere above 100 km contain charged particles, and below that level almost entirely neutral ones? And with respect to composition, why does dry air to heights up to about 100 km contain the same proportions of carbon dioxide, argon, oxygen, and nitrogen, even though the molecular weights are 44, 40, 32, and 28, respectively, and we should expect the heavier gases to be separated from the lighter gases by gravity? Within this uniformity why is there a layer in which oxygen takes the form $O_3$ (ozone) rather than $O_2$ (molecular oxygen)? And why is the layer of a relatively high (but still very low in absolute magnitude) concentration of ozone located in the upper stratosphere, in spite of the fact that its molecular weight (48) is the highest of common atmospheric gases?

The answer to all of these questions lies entirely or in large part with the disposition of radiant energy from the sun as it passes through the atmosphere to the ground. The energy that heats the atmosphere and drives its circulations originates in the sun. Small amounts of energy from other stars reach the earth, and minute quantities are released by radioactive transformations of matter on earth, but these are extremely small indeed compared with the large steady flow of energy from the sun.

The next chapter deals with the nature of radiation, the properties of solar radiation, and the interaction of solar radiation with the atmosphere. In it are presented explanations of some of the characteristics of the atmospheric structure that were described in this chapter.

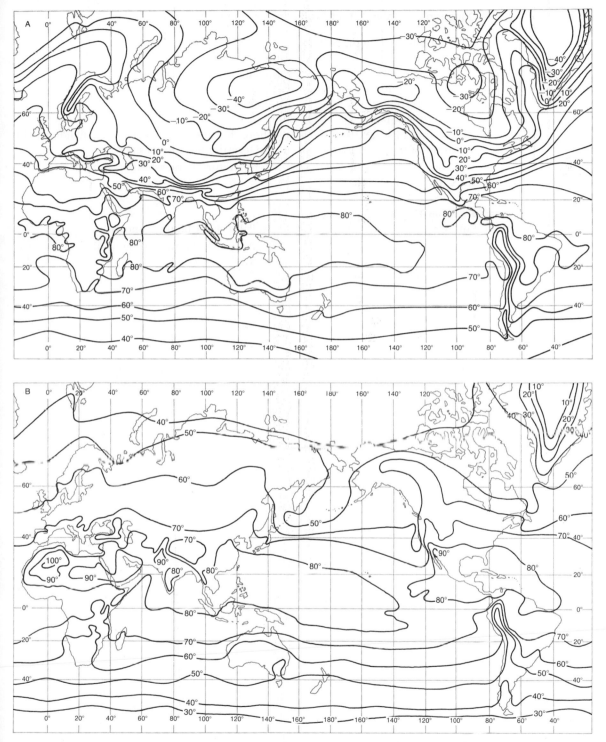

**Figure 2.7** *Average surface air temperatures over the earth (°F): (A) January; (B) July. [From A. N. Strahler, Physical Geography, 3d ed. (New York: Wiley, 1969).]*

Questions, Problems, and Projects for Chapter 2

1  Define the following terms:

| | | | |
|---|---|---|---|
| a | Atmosphere | f | Atom |
| b | Troposphere | g | Molecule |
| c | Stratosphere | h | Element |
| d | Mesosphere | i | Compound |
| e | Thermosphere | j | Ion |

2  List the constituents of the atmosphere that affect the weather in a major way in the order of their relative importance. Describe the role of each.

3  From Figure 2.6 make graphs of temperature versus height at 5°N, 45°N, and 75°N in summer and winter, and compare the resulting curves with those in Figures 2.4 and 2.5.

4  From Figure 2.7 make graphs of surface air temperature versus latitude at the following longitudes: 0°, 80°W, 180°, and 100°E in January and July.

5  Radio waves, like light, normally travel in straight lines. In view of the curvature of the earth's surface, why can radio signals be received at great distances from the place at which they are transmitted?

# THREE

## Radiation Through the Atmosphere

### 3.1 The sun and the earth

The earth revolves around the sun at an average distance of $149.6 \cdot 10^6$ km (about 150 million kilometers or 93 million miles).[1] Its orbit is an ellipse, but when it is closest to the sun (the perihelion), on about January 4, it is $147 \cdot 10^6$ km away, only 1.7 percent nearer the sun, and at its aphelion, about July 5, it is the same percentage farther away. Thus its path around the sun departs only a little from a circle. At all times of the year the energy that controls the temperature and motions of the atmosphere must travel about 150 million kilometers, mostly through empty space, to reach the earth. How does it do it?

Another puzzling question concerns the amount of energy the sun is sending out. The sun is not "beaming" its energy only towards the earth; it sends it out equally in all directions. The earth intercepts only a small fraction of it, although this amount is $170 \cdot 10^{12}$ kilowatts, or more than 500,000 times the capacity of all the electric-generating plants in the United States. This huge amount of energy, which keeps the earth warm, provides for the growth of plants and the life of animals and man, and creates the circulation of the atmosphere, is an almost negligible part of the solar energy. The total energy emitted by the sun is more than 2 billion

---

[1]See Appendix A for an explanation of the "power of ten" notation.

times as much as the earth receives. If this tremendous amount of energy were produced by combustion, the sun would have burned up long ago; an amount of coal equal to the mass of the sun could produce energy at this rate for less than five thousand years. Where does all this energy come from?

The answer to the second question is that the sun is a huge nuclear reactor. At the extremely high temperatures and pressures in the interior of the sun, a hydrogen-helium fusion process takes place by which mass is transformed into energy. The amount of mass used in this process is so small that it will take about $14 \cdot 10^{12}$ years for the sun's mass to be exhausted. The estimated age of the solar system is $4.6 \cdot 10^9$ years, so at the present rate of emission of energy the sun can be expected to last more than 3,000 times as long at it has existed so far.

How this emitted energy is transferred through empty space is more difficult to answer. It is a question of the same fundamental nature as the question of how the gravitational attraction between two masses is communicated through empty space. For our purposes it will suffice to say that energy transfer does take place without the requirement of a medium to transmit it and that this form of energy transfer is called *radiation*.

## 3.2 The nature and laws of electromagnetic radiation

Radiation is a term used both for a process of propagation of energy through space and for the energy that flows in this manner. It is of two types: electromagnetic wave radiation, of which radio waves and light are familiar types, and corpuscular radiation, which includes alpha rays (beams of helium nuclei), and beta rays and cathode rays (beams of electrons), as well as the neutron, proton, and meson beams of modern high-energy physics.

Corpuscular radiation from the sun accounts for only one-millionth of the total solar energy incident on the earth and its extended atmosphere. Nevertheless, under certain situations the bombardment of our atmosphere by corpuscular beams is the dominant means of heating and ionizing the upper atmosphere. These effects are most pronounced in the polar regions where they produce the magnificently varied forms of the *aurora borealis* (sometimes referred to as the Northern Lights) and the *aurora australis* (Southern Lights). Our understanding of the fascinating spectacle of the aurora is still very limited, despite the fact that it has been studied for well over twenty-five centuries. The present view is that it is caused by a focusing effect of the earth's geomagnetic field on moving charged particles. The magnetic forces, which have influence to a distance approximately ten times the earth's radius, control the motion of the charged particles in the radiation belts. Particles trapped in the outer reaches of our atmosphere (the magnetosphere) are known to be precipitated period-

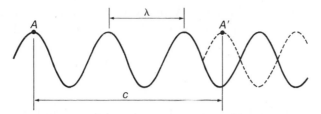

**Figure 3.1** *Schematic representation of the characteristics of a moving wave: the distance from one crest to the next is the wavelength λ; the dashed curve at A' is the position of the wave crest initially at A after unit time (for instance, one second); and c is the distance traveled in unit time, that is, the wave speed.*

ically into the lower atmosphere during disturbed periods that we call geomagnetic storms. The physical process responsible for this transfer of energy from high levels to lower levels in the atmosphere is still uncertain, but it is thought to involve a large-scale *collapse* of the entire outer magnetosphere. As the precipitating charged particles descend to levels where the density is higher, they collide with neutral atoms or molecules and either raise them to an excited state (of increased energy of their electrons) or remove one of their orbital electrons, thereby ionizing them. The subsequent decay of the excited particles to their ground states involves emission of quanta of electromagnetic radiation that is responsible for the characteristic red, green, and violet colors of active aurora.

Most of the energy incident on the earth, however, is in the form of solar *electromagnetic waves*. It is the behavior of this type of radiation that is largely responsible for the variation of temperature with height in the various layers of the atmosphere.

The waves we are most directly familiar with are those occurring at the surface of a body of water, for instance, ocean waves. In general, these are very irregular and complex, but their main properties are observed readily, particularly near shore where the waves are simpler. They consist of a series of moving troughs and ridges in the water surface (Figure 3.1). The horizontal movement of the waves is not produced by bodily horizontal movement of the water, but primarily by alternating up and down movements, up ahead of the crest and down behind the crest and ahead of the trough. The *wave speed* is thus different from the speed of flow of the water. It is the speed with which an identifiable part of the wave, for instance, the crest, moves. In Figure 3.1 the dashed curve marked A' represents the wave crest originally shown by the solid curve at A after a unit time interval. The wave speed is thus equal to the distance AA' per unit time. The distance from corresponding parts of successive waves at a given time (e.g., from crest to crest) is called the *wavelength*. The number of waves that go by a point in unit time is called the *wave frequency*, and the time that it takes for a complete single wave to pass is called the *wave period.* Since the waves between *A* and *A'* in the solid curve in the diagram would pass *A'* in unit time, their number (approximately 2.5) is the frequency. If this number is represented by $\nu$, the time $\tau$ for one of them to pass is $1/\nu$, that is, the period is the reciprocal of the frequency $\nu$:[2]

$$\tau = 1/\nu \tag{3.1}$$

The wavelength and wave speed are also interrelated. If we represent the wave speed by *c* and its length by $\lambda$,

$$c = \lambda/\tau = \lambda\nu \tag{3.2}$$

As a wave moves it transfers energy (the ability to do work) from place to place. The lifting of a boat by the crest of a water wave as it moves by demonstrates this ability to do work, and the lifting in succession of a group of small boats at various distances from a passing ship demonstrates the transfer of energy by the bow wave of the ship.

Electromagnetic waves differ from water waves in that they do not require the presence of matter to transport energy. At one time physicists were puzzled by the propagation of waves through a vacuum. Since it was difficult to conceive of waves without some medium doing the waving, they postulated the existence of an all-pervading substance, the "luminescent ether." As the nature of forces operating at a distance, such as the force of gravity and electrostatic and electromagnetic forces, became better understood, the idea of the oscillation of a force at a point in space owing to the vibrating movement of a distant electric charge became more acceptable, even if there is no matter at the position where the oscillation takes place.

Electromagnetic waves move with a speed of $3 \cdot 10^8$ m s$^{-1}$ (186,000 miles per second) in empty space and only slightly slower in space that is occupied by air. Their wavelengths range from several kilometers for long radio waves to billionths of a centimeter or less for X rays and gamma rays. Figure 3.2 shows the ranges of wavelengths and the corresponding frequencies for various kinds of electromagnetic radiation. The wavelengths of the radiation that carries most of the energy from the sun to the earth's atmosphere and from the earth and its atmosphere to space are in the vicinity of those of visible light. These wavelengths are so short that it is convenient to express them in microns, where 1 micron (1 $\mu$m) $= 10^{-6}$ m $= 10^{-4}$ cm, or in Ångstrom units, where 1 Å $= 10^{-8}$ cm.

In these units visible light ranges from 0.4 $\mu$m or 4000 Å (violet light) to 0.7 $\mu$m or 7000 Å (red light).[3] Radiation with wavelengths shorter than 0.4 $\mu$m is called *ultraviolet*, and that with wavelengths longer than 0.7 $\mu$m is called *infrared*.

---

[2] See Appendix C for names of Greek letters.

[3] The human eye is actually sensitive to light of wavelengths slightly shorter and longer than this range, but so slightly that these limits are commonly used.

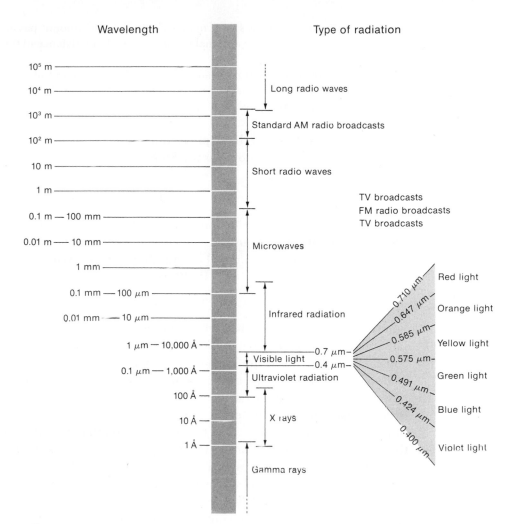

**Figure 3.2** *The electromagnetic spectrum.*

The behavior of electromagnetic radiation is described by certain physical laws. The following is a simplified summary of them.

1 All matter that is not at absolute zero can emit radiation in amounts that depend on its temperature.

2 Some substances emit radiation only in certain wavelengths. This is particularly true of gases.

3 Substances that emit the maximum amount for their temperature in all wavelengths are called *black bodies*. The amount of radiation emitted by a black body is proportional to the fourth power of its absolute temperature.

4 Substances absorb only radiation of wavelengths that they can emit.

5 The hotter the substance, the shorter the wavelengths at which most of the radiant energy is emitted.

6 Since the total radiation from a point source passes through the surface of

successively larger spheres as it radiates outward, the amount passing through a unit area is inversely proportional to the square of the distance of the area from the source.

7 If a gas absorbs radiation of any wavelength, the amount absorbed will be proportional to (a) the number of molecules of gas and (b) the intensity of radiation of that wavelength.

Some of these laws can be expressed mathematically very simply. Thus law number 3, which is known as the Stefan-Boltzmann law, is represented by the equation

$$E = \sigma\, T^4 \qquad (3.3)$$

where $E$ is the energy emitted per second from unit area of a black body whose temperature is $T$, and $\sigma$ is a constant that has the value $5.67 \cdot 10^{-8}$ watts/m$^2$ or $5.67 \cdot 10^{-5}$ ergs/cm$^2$. Law number 5, called Wien's displacement law (after Wilhelm Wien, 1864–1928), is

$$\lambda_M = a/T \qquad (3.4)$$

where $\lambda_M$ is the wavelength at which the peak occurs in the spectrum of the radiation from a black body whose absolute temperature is $T$, and $a$ is a constant that has the value 2898 if $\lambda_M$ is expressed in microns.

Both of these laws are consequences of a more general law, Planck's radiation law, which gives the complete distribution of radiant energy from a black body at different wavelengths. Planck's law is of special importance because in his effort to find a formula that gave the distribution that was observed Max Planck (1858–1947) found it necessary to postulate that radiant energy was emitted in discrete packets, which he called *quanta*. Previously, Wien had derived an empirical expression on continuous radiation theory that fit the experimental data for short wavelengths and gave the shift of the maximum towards shorter wavelengths when the temperature goes up as shown in his displacement law, but did not correspond to experimental results at longer wavelengths. Lord Rayleigh and Sir James Jeans (1877–1946) derived an equation based on the theory of statistical mechanics that, while apparently based on sound theoretical considerations, fit the observed data only at very long wavelengths. In evolving a theory that matched the experimental results for all wavelengths, Planck introduced the *quantum theory*, which has completely revolutionized physics.

Figure 3.3 shows the radiation spectrums for the emission from a black body at (A) 6000°K, approximately the effective temperature of the sun's surface, and (B) some temperatures representative of the range occurring at the surface of the earth. Note that the wavelength of the energy maximum for 6000°K is near 0.5 $\mu$m, in the visible range, while for terrestrial temperatures it is in the far infrared, at about 10 $\mu$m. Note also that at all wavelengths the energy emitted at higher temperatures is higher than at lower temperatures; even the peak emission for terrestrial temperatures is

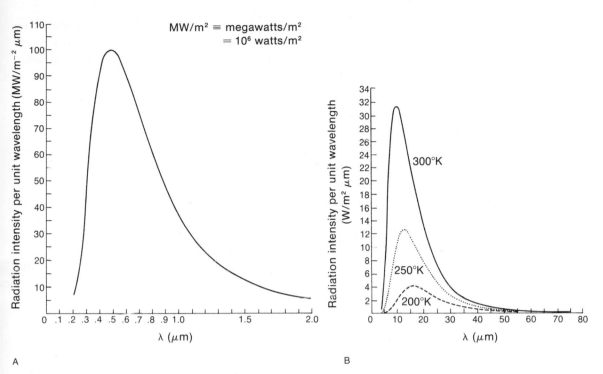

**Figure 3.3** *Variation of intensity of black-body radiation with wavelength: (A) T = 6000° K; (B) T = 200°, 250°, and 300° K.*

only a very small fraction of that at 6000°K at the same wavelength. The effect of distance from the sun to the earth, as expressed by law number 6, just offsets the temperature effect and permits the radiation received from the sun and emitted by the earth to be in balance. This law may be written as follows:

$$E = C/d^2 \tag{3.5}$$

or

$$E_1/E_2 = d_2^2/d_1^2$$

where $E$ is the flux of radiation (i.e., the radiant energy passing through unit area per second) at a distance $d$ from the source; $E_1$ is the flux of radiation at a distance $d_1$ from the source; $E_2$ is the flux of radiation at a distance $d_2$ from the source; and $C$ is a constant depending on the intensity of the source and representing the radiant energy at unit distance. We shall see how these laws are applied when we discuss the radiation from the sun in Section 3.4.

## 3.3 Energy, heat, and work

The word "energy" is part of our everyday language, and in a general way it is commonly understood. Unless you have studied its nature in a course in physical sciences, however, its precise nature and, in particular, the units in which its measured values are expressed may not be familiar to you.

Energy is defined as the capacity for doing work. This capacity may take various forms, some of which follow:

*Potential energy:* Energy due to position, such as that produced by lifting a weight. Potential energy can be released by letting go of the weight and allowing it to fall.

*Kinetic energy:* Energy due to motion, for instance, the motion of a hammer swinging toward the head of a nail.

*Heat:* Energy associated with the ability of one body to raise the temperature of a cooler one. The capacity of heat to perform work is illustrated, for instance, in the steam engine.

*Chemical energy:* Energy stored by the combination of molecules into chemical compounds. Examples of the release of chemical energy are the combustion of gasoline that propels automobiles and the chemical reactions inside dry cells (batteries) that produce electrical energy.

*Nuclear energy:* Energy stored in the nucleus of an atom.

*Radiant energy:* Energy associated with electromagnetic waves propagating through space, whether empty or occupied by a substance. If radiant energy can pass through a substance without change, the substance is said to be *transparent* for radiation of that wavelength.

One form of energy can be transformed into another. Thus the release of a suspended weight results in the transformation of its potential energy into kinetic energy. The burning of gasoline in the cylinders of an automobile results first in the conversion of chemical energy into heat and then of part of that heat into the kinetic energy of the automobile's motion. The absorption of radiant energy passing through air may heat it; the part that passes through the air and reaches the ground heats the ground.

The equivalence of heat and work or mechanical energy was first demonstrated by Benjamin Thompson, Count Rumford (1753–1814), toward the end of the eighteenth century. Rumford, who was born near Boston but sympathized with the British and went to England at the time of the American Revolution, had a very impressive career both in England and in Austria and Bavaria. He was noted for his introduction of important social reforms in Bavaria, but his most important contribution was the demonstration that the then-current view that heat was a material fluid, called caloric, was wrong. He showed this in a series of experiments he was led to by considering the heat arising from friction during the boring of cannons at the military arsenal in Munich of which he was in charge.

Even now different units are frequently used to express heat energy — the calorie in the metric system and the BTU (British Thermal Unit) in the English system of units — from those used for other forms of energy. One calorie is the amount of energy required to raise the temperature of one gram of water one degree Celsius (from 14.5°C to 15.5°C), and one BTU is that required to raise the temperature of one pound of water one degree Fahrenheit (from 62°F to 63°F). [The calorie in which energy produced by food in the body is expressed is the "large calorie" or kilogram calorie, 1,000 times as large as the (gram) calorie.]

Since all forms of energy represent the ability to do work, the natural way in which to express quantity of energy is in work units. In physical science there are two systems of units in common use. The cgs (centimeter-gram-second) system has been replaced recently by the mks (meter-kilo-gram-second) system as the International System, but many scientists still use the cgs system. The unit of energy in the cgs system is the erg, which is the amount of work done when a force of unit magnitude in this system (one dyne) acts through a distance of one centimeter. The joule, the unit of energy in the mks system, is similarly the work done when unit force in this system (one newton) acts through one meter. One joule is equal to $10^7$ (10 million) ergs. The rate at which energy is released, or work is done, is expressed in watts in the mks system, with one watt equal to one joule per second. One calorie is equal to 4.186 joules or $4.186 \cdot 10^7$ ergs.

In the English system the unit of work energy is the foot-pound, and the rate of doing work is expressed in horsepower, where one horsepower equals 550 foot-pounds per second.

In discussions of radiation we are frequently concerned with the amount of energy falling on or passing through a unit area. In the cgs and mks units this is expressed in ergs/cm² and J/m², where 1 J/m² = $10^3$ ergs/cm². The unit for this quantity in heat units has been given a special name, the langley (ly), named after Samuel P. Langley (1834–1906), the American scientist who, in addition to his experiments in heavier-than-air flight, carried on important studies of radiation. One langley is one calorie per square centimeter; thus 1 ly = $4.186 \cdot 10^7$ ergs/cm² = $4.186 \cdot 10^4$ J/m².

The rate at which the energy falls on or passes through unit area is called *radiant flux density* or *irradiance*. It is usually expressed in watts per square meter (sometimes per square centimeter) or langleys per minute, where 1 ly/min = $6.98 \cdot 10^2$ W/m².

## 3.4    Solar radiation

Except for the effect of distance, as stated in law number 6 of Section 3.2, the radiation from the sun reaches the outer limit of the earth's atmosphere undepleted, but in passing through the air it is scattered and absorbed by the molecules and the dust, haze, and cloud particles, so that only a part of it reaches the earth's surface. Consequently, previous to the development of rockets and artificial satellites, the amount of radiation coming from the sun could only be estimated from the measurements made at

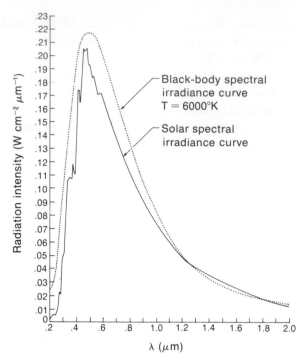

**Figure 3.4**  *Solar irradiance spectrum and 6000°K black-body radiation reduced to mean solar distance.*

observatories on high mountains. Even today reliable measurements from satellites are not available for the incoming radiation from the sun, and the best values are those obtained many years ago by Abbott, Fowle, and other staff members of the Smithsonian Institution from a long series of measurements at Mt. Wilson, corrected and augmented by rocket measurements of ultraviolet radiation, which is almost completely blocked out by absorption in the upper atmosphere.

These measurements, extrapolated to the outer limits of the atmosphere and corrected for mean solar distance, are shown in Figure 3.4. The figure shows the amount of energy of various wavelengths received on a unit area perpendicular to the line from sun to earth per second. The total of this energy in all wavelengths, which is called the *solar constant*, is 1.94 calories per square centimeter per minute, or 1.35 kilowatts per square meter.

The solar constant is the radiant flux before depletion by the atmosphere at the mean distance of the earth from the sun, that is, $149.6 \cdot 10^6$ km. From its value we can use the inverse square law to compute the rate at which radiant energy is emitted at the sun's surface. The radius of the visible surface of the sun is $6.95 \cdot 10^5$ km. If we call this distance from the sun's center $d_1$, and the mean solar distance $d_2$, the solar constant is $E_2$ and the rate at the sun's surface, which we wish to calculate, is $E_1$. Putting the numbers in equation (3.5), we obtain

$$E_1 = \frac{(149.6 \cdot 10^6)^2}{(6.95 \cdot 10^5)^2} \cdot 1.35 \text{ kW/m}^2$$

$$E_1 = 6.26 \cdot 10^4 \text{ kW/m}^2 = 6.26 \cdot 10^7 \text{ W/m}^2$$

If now we apply equation (3.3) to obtain the effective temperature of the sun's surface, we have

$$6.26 \cdot 10^7 = 5.67 \cdot 10^{-8} \, T^4$$

$$T = \sqrt[4]{\frac{6.26 \cdot 10^7}{5.67 \cdot 10^{-8}}} = \sqrt[4]{1106 \cdot 10^{12}}$$

A convenient way of obtaining the fourth root of a number is by the use of logarithms. Doing so, we find

$$T = 5770°\text{K}$$

Thus the total amount of radiation emitted by the sun is approximately that of a black body at a temperature of 5770°K.

The wavelength distribution of solar radiation is also similar to that of a black body at 5770°K. In Figure 3.4, in addition to the curve based on measurements of solar radiation, the curve for the radiation from a black body at 6000°K reduced to the mean distance from sun to earth is shown. (We used 6000 rather than 5770 because the data for it are available in published tables.) The peaks in the two curves are at about the same wavelength (0.47 $\mu$m for the observed, 0.50 $\mu$m for the 6000°K curve) and the general shape is much the same, although the measured solar radiation is slightly deficient in the ultraviolet and slightly excessive in the near infrared. On the short wavelength end the observed curve falls almost to zero at 0.2 $\mu$m, and on the long wavelength end it reaches very low values by 2.0 $\mu$m. Practically all the energy is in wavelengths between 0.15 and 4.0 $\mu$m. About 9 percent is in the ultraviolet, 41 percent in the visible, and 50 percent in the infrared.

The solar constant is defined as the radiation that falls on unit area of surface normal to the line from the sun per unit time at the outside of the atmosphere and at *mean solar distance*. The deviation from this value due to varying distance from the sun is slight. The 1.7 percent departure in distance leads to about 3.4 percent difference in radiation. That is, on January 4 there is 3.4 percent more, or 1.40 kW/m², falling on a unit area normal to the line from the sun at the outside of the atmosphere, and about July 5 the amount is 1.30 kW/m². The big variations in energy from the sun reaching the ground at various times of the year are not due to this effect but are due to the inclination of the earth's surface to the sun and the depletion of the energy from the sun in passing through the atmosphere.

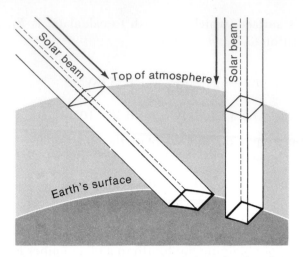

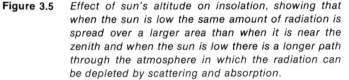

**Figure 3.5** *Effect of sun's altitude on insolation, showing that when the sun is low the same amount of radiation is spread over a larger area than when it is near the zenith and when the sun is low there is a longer path through the atmosphere in which the radiation can be depleted by scattering and absorption.*

## 3.5 Insolation at the earth's surface

The amount of solar radiation received on a *horizontal* surface is called *insolation*. Insolation usually refers to the solar radiation received at the earth's surface per unit area and unit time, and we shall use it in this sense. It includes both the direct radiation from the sun and the indirect radiation from the sky due to scattering of the solar radiation by molecules or other particles. (If it weren't for this scattering the sky would be black during the day as well as at night.) The term "global radiation" is sometimes used instead of insolation.

The insolation varies because of the variation of solar altitude for two reasons: (1) When the sun is high the radiation falls almost perpendicularly on the ground, whereas when it is low the radiation passing through unit area normal to the line from the sun is spread over a large area (see Figure 3.5). (2) When the sun is low the radiation traverses a longer path through the atmosphere along which a greater amount of scattering and absorption can take place. Both effects reduce the insolation when the sun is low, in accord with our common experience during any clear day. At midday the sun's radiation is much more intense than in the early morning and late afternoon.

The total insolation for a whole day is affected by the length of time from sunrise to sunset, in addition to the above two factors.

The insolation varies diurnally, with time of year, and with latitude because of these factors. The diurnal variation, already mentioned, is an obvious consequence of the changing height of the sun during the day.

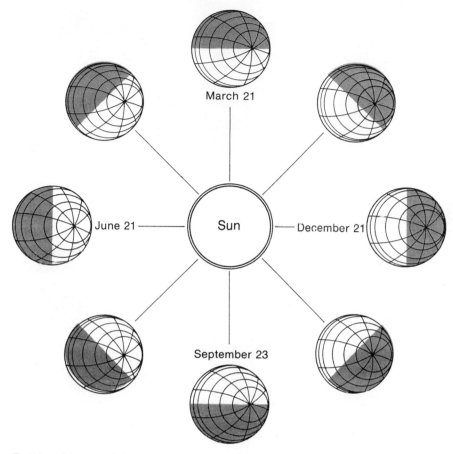

March 21

June 21 —— Sun —— December 21

September 23

**Figure 3.6** *Earth's orbit around the sun.*

The seasonal variation is a consequence of the inclination of the earth's axis to the plane of the ecliptic (the plane in which its orbit around the sun lies). The direction of the earth's axis of rotation stays practically the same throughout the motion of the earth around the sun, maintaining an angle of 66½° with the orbital plane. The consequence in terms of the length of day and height of the sun at noon may be seen in Figures 3.6 and 3.7. As shown in Figure 3.6, while the earth goes around the sun with the axis pointing in a constant direction, the area that is illuminated by the sun changes; it includes the North Pole at the June solstice and excludes it in December. Figure 3.7 shows this effect in more detail. The fraction of the entire day during which the sun shines at each latitude is shown by the portion of the parallel of latitude that is in the illuminated hemisphere. At the June solstice the sun shines throughout the entire period of rotation of the earth at all latitudes north of the Arctic Circle, while south of the Antarctic Circle there are 24 hours of darkness. In the latitudes between the Arctic and Antarctic Circles there are progressively fewer hours of daylight as one goes from north to south. At the equator there are 12 hours

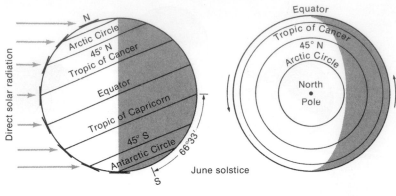

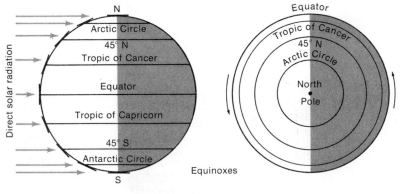

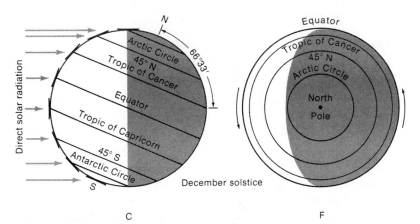

**Figure 3.7** *Exposure of the earth to the sun's radiation: A, B, and C show the attitude of the midday sun; D, E, and F, the relative length of daily periods of sunshine and darkness.* [*From "Seasons," by John B. Leighly,* Encyclopaedia Britannica, © *1962.*]

**Table 3.1**    Time from Sunrise to Sunset

| Latitude | Winter solstice | Vernal or autumnal equinox | Summer solstice |
|---|---|---|---|
| 90° | 0 | 12 hr 0 min | 6 months |
| 80° | 0 | 12 hr 0 min | 4 months |
| 70° | 0 | 12 hr 0 min | 2 months |
| 60° | 5 hr 33 min | 12 hr 0 min | 18 hr 27 min |
| 50° | 7 hr 42 min | 12 hr 0 min | 16 hr 18 min |
| 40° | 9 hr  8 min | 12 hr 0 min | 14 hr 52 min |
| 30° | 10 hr  4 min | 12 hr 0 min | 13 hr 56 min |
| 20° | 10 hr 48 min | 12 hr 0 min | 13 hr 12 min |
| 10° | 11 hr 25 min | 12 hr 0 min | 12 hr 38 min |
| 0° | 12 hr  0 min | 12 hr 0 min | 12 hr  0 min |

of daylight and 12 hours of darkness. At the December solstice the duration of daylight is just reversed, with the region south of the Antarctic Circle having 24 hours of daylight. At the equinoxes there are 12 hours of daylight at all latitudes.[4]

The variation of length of day with season and latitude is shown in Table 3.1. At the equator it is 12 hours throughout the year, but at 60° latitude it ranges from less than 6 hours in winter to more than 18 in summer, and of course at the poles the sun is below the horizon for six months and above the horizon for the other six months.

Figure 3.7 also shows how the height of the sun at noon varies with latitude and season. At the June solstice the noon sun is at the zenith (exactly overhead) at the Tropic of Cancer; farther north it is lower, but even at the North Pole it makes an angle of 23½° with the horizontal plane (the solar altitude) and 66½° with the vertical (its zenith angle). As one goes south the noon sun is lower, and at the Antarctic Circle it is at the horizon. Conditions are just reversed at the December solstice, with the noon sun overhead at the Tropic of Capricorn and on the horizon at the Arctic Circle. At the equator the noon sun is at the zenith at the equinoxes and departs at most 23½° from it at the solstices, so that the sun's radiation is always almost perpendicular to the horizontal at noon. At the poles the highest the sun gets above the horizon is 23½°, so that even when it is above the horizon 24 hours a day the radiation spreads over a large horizontal area all day.

Figure 3.8 shows the distribution of daily insolation that would be received at the earth's surface at various latitudes and seasons if there were no atmosphere. It is seen that at the equator, where the noon sun is never far from the zenith and the sun is above the horizon 12 hours every day, the total insolation before depletion by the atmosphere varies only about

[4]Actually, the time from sunrise to sunset is slightly longer than indicated by these astronomical considerations because the sun's rays are refracted (bent) by the atmosphere, making the sun appear to be at the horizon when it is slightly below it; daylight is also increased by twilight, which is produced by the scattering of sunlight by the upper part of the atmosphere, which remains illuminated after the sun has set for an observer at the ground.

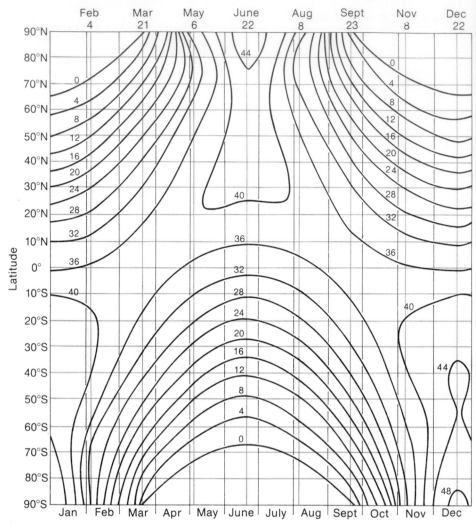

**Figure 3.8**  *Solar radiation (MJ/m² day) falling on horizontal surface at outside of atmosphere.*

12 percent, from a high of about 38 MJ/m² day (megajoules/m² day) in March to a low of about 32 MJ/m² day in June. At 45°N latitude it ranges from about 12 MJ/m² day at the winter solstice to about 41 MJ/m² day at the summer solstice, and poleward of 66½° it varies from zero at the winter solstice to more than 42 MJ/m² day at the summer solstice.

The small added effect of the varying distance from the sun gives the South Pole the largest amount of radiation on a horizontal surface outside of the atmosphere in a day anywhere on earth, 48 MJ/m² on December 22. The maximum at the North Pole is 45 MJ/m² on June 21. It should be noted that the low angle of the sun tends to offset the duration of sunlight at the poles, but, even so, in the absence of the atmosphere the sun would heat the poles more than the equator at the summer solstice. That the poles are cooler than lower latitudes in summer is due to the depletion of the radia-

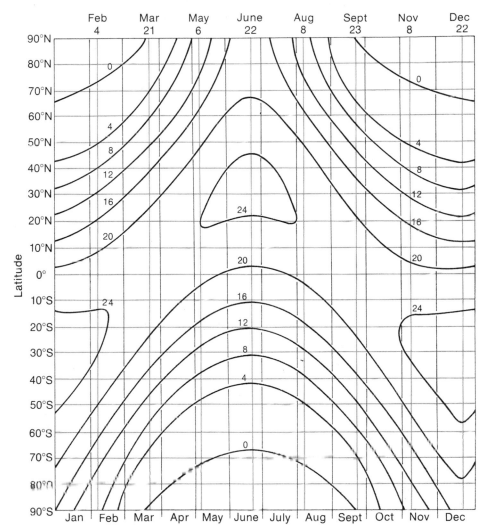

**Figure 3.9**    *Solar radiation (MJ/m² day) falling on horizontal surface at sea level, assuming atmospheric transmission of 0.7 for vertical sun and cloudless skies. [Milankovich's computed values, adjusted to presently accepted value of solar constant.]*

tion as it passes through the air, as well as such effects as the reflection of radiation by snow and ice and the utilization of heat in melting the snow and ice.

In the next section we shall discuss how radiation of different wavelengths is depleted differently by the constituents of the atmosphere. To get a rough idea of how the insolation is affected by the differences in path length through the air at different latitudes and seasons, Milankovich computed the radiation on the assumption that if the sun were in the vertical, 70 percent of the incident radiation would be transmitted to the ground everywhere. The results of his computations, adjusted for the presently accepted value of the solar constant, are shown in Figure 3.9.

In this diagram we see that the maximum insolation is near the latitude at which the sun is vertical at noon. At the summer solstice in each hemisphere the amount varies little from pole to equator. The maximum then is at about 30° latitude, totaling more than 25 MJ/m² day in the Southern Hemisphere and a little less in the Northern Hemisphere. The equator receives about 20 MJ/m² day and the poles both receive somewhat more than 16 MJ/m² day at their respective summer solstices. At the winter solstice the variation of daily insolation with latitude is large, from zero in the region of polar night to about 20 MJ/m² day at the equator.

The assumed transmissivity of 70 percent is frequently exceeded, so that daily insolation amounts in excess of 30 MJ/m² (720 langleys) even at low altitude stations are not rare, and mountain stations may have even higher values. On cloudy days the insolation is greatly reduced.

Of the insolation falling on the earth's surface some is absorbed and some is reflected back towards the sky. The fraction of the solar radiation that is reflected is called the *albedo*. At the earth's surface the albedo depends on the character of the surface, and at water surfaces it depends also on the angle of the sun above the horizon. Table 3.2 gives approximate values of the albedo for various types of surfaces, for clouds, and for the earth and atmosphere as a whole (the planetary albedo).

## 3.6  Depletion of solar radiation by the atmosphere

Radiation passing through the air is subjected to two types of attenuation: *scattering* and *absorption*.

When radiation is scattered by a particle, say, an air molecule, it is as though the radiation were momentarily captured and then sent out unchanged in amount and wavelength, but dispersed in all directions, not in just the direction from which it came. Consequently, of the radiation from the sun thus captured only about one-half continues downward, the other half being sent back into space. Radiation scattered by air molecules comes from all parts of the sky, not just from the direction of the sun; thus the daytime sky appears bright. Small particles like molecules scatter a larger proportion of the short-wave radiation, the blue and violet light, than the longer-wave yellow and red light. That is the reason the sky is blue in the absence of haze or smog. Haze, fog, and smog contain larger particles, which scatter more nearly equally in all wavelengths. When they are present, the sky tends to be white, particularly near the horizon, except when smog with absorptive properties gives it a yellow or brownish color.

In absorption, air molecules actually take up part of the radiant energy and convert it into internal energy, whereby the motion of the molecules or their component atoms and electrons is changed. This energy increase manifests itself as a temperature change. Layers in which radiation is absorbed may be expected to be the warm layers of the atmosphere. Let us consider what happens to solar radiation as it passes downward.

**Table 3.2**  Albedo of Various Surfaces

| Surface | % |
| --- | --- |
| Fresh snow | 80–85 |
| Old snow | 50–60 |
| Sand | 20–30 |
| Grass | 20–25 |
| Dry earth | 15–25 |
| Wet earth | 10 |
| Forest | 5–10 |
| Water (sun near horizon) | 50–80 |
| Water (sun near zenith) | 3–5 |
| Thick cloud | 70–80 |
| Thin cloud | 25–50 |
| Earth and atmosphere | 35 |

The components of the air, like all gases, absorb radiation of certain wavelengths and are transparent to others. Nitrogen, whether molecular or atomic, absorbs only in the far (very short-wave) ultraviolet, of which there is very little in the sun's radiation. This very short-wave radiation, when absorbed by a molecule or atom, can knock a molecule into separate atoms or an electron out of an atom. The small amount of very short-wave ultraviolet radiation available is used up at very great heights; at these heights it dissociates molecular nitrogen, ionizes the atoms, and heats the very rarefied air.

Oxygen absorbs more strongly than nitrogen and over a wider range of wavelengths in the ultraviolet. It is thus also dissociated and ionized in the upper part of the atmosphere. By the time the radiation reaches 100 km, all of the far ultraviolet has been absorbed. Below this level is a region in which there is little absorption because none of the radiation that is strongly absorbed reaches it, and the density (number of molecules) is insufficient for weakly absorbed radiation to be important. This region extends for some distance above and below 80 km. The lack of absorption of radiation causes the minimum of temperature there.

At still lower levels an interesting development takes place. The weak absorption by oxygen of the radiation that reaches these levels produces some atoms, but now the density of molecules is great enough for the atoms to collide with the ordinary diatomic molecules and produce triatomic molecular oxygen (ozone). The ozone absorbs ultraviolet radiation of longer wavelengths (up to about 0.3 $\mu$m), and since there is more of this radiation, both because the intensity at these wavelengths incident on the atmosphere is greater and because it has not been subject to absorption higher up, it heats up the layer to quite high temperatures, producing the temperature maximum at 50 km.

At levels below 50 km the ultraviolet radiation that ozone can absorb gets used up, so that with the increased density of air to heat and the decreased absorption the temperature is lower.

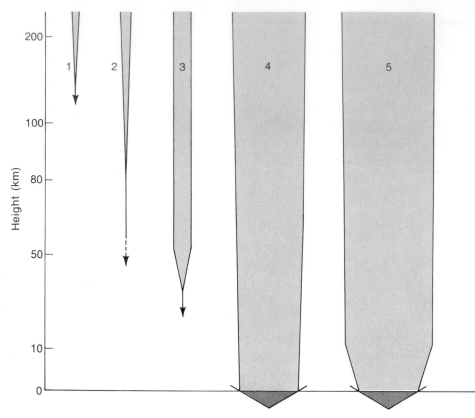

**Figure 3.10** *Transfer and absorption of solar radiation.*
*1. UV, $\lambda < 0.12$ $\mu m$, absorbed by $N_2$ and $O_2$.*
*2. UV, $0.12$ $\mu m \leq \lambda < 0.18$ $\mu m$, absorbed by $O_2$.*
*3. UV, $0.18$ $\mu m \leq \lambda < 0.34$ $\mu m$, absorbed by $O_3$.*
*4. Near UV and visible, $0.34$ $\mu m \leq \lambda < 0.7$ $\mu m$, transmitted nearly undiminished except for scattering.*
*5. Near IR, $0.7$ $\mu m \leq \lambda < 3$ $\mu m$, absorbed slightly by $O_2$, and in troposphere by $H_2O$ vapor.*

The solar radiation of wavelengths greater than 0.3 $\mu$m is absorbed only weakly or not at all by the atmosphere. While there is some absorption of the visible and infrared radiation, particularly at very low levels where the water-vapor content is large, the amount of insolation absorbed in the lower stratosphere and troposphere is very small and effectively these layers may be regarded as transparent to solar radiation. Consequently, except for radiation of wavelengths shorter than 0.3 $\mu$m, which is absorbed at high levels by oxygen and ozone, most of the sun's radiation comes through the atmosphere and is absorbed by the ground. The temperature of the ground is raised by absorbing the radiation.

The disposition of solar radiation in passing through the atmosphere is illustrated schematically in Figure 3.10. The various wavelengths of incoming radiation are represented by bands terminating in arrowheads, with the width of the bands representing crudely the amount of incident radiation in those wavelengths. The depletion by absorption is represented by the narrowing of the bands, and the arrowhead shows the depth in the atmosphere to which the radiation of those wavelengths penetrates. The very short ultraviolet radiation is absorbed by nitrogen and oxygen at the levels of the F-layers of the ionosphere. The slightly longer wavelength radiation that dissociates oxygen is mostly absorbed above 80 km, but a little at the long wavelength end of this band penetrates below 50 km. The ultraviolet that is absorbed by ozone begins being absorbed below 80 km, and is practically gone by the time 20 km is reached. The rest of the radiation, on the other hand, almost all gets through to the ground, although some, in the infrared, is absorbed by water vapor.

Where the radiation strikes solid ground, the part that is not reflected is converted to heat right at the surface, and this heat raises the temperature of a very thin layer of soil or rock; as a consequence the temperature increase is large. Where the earth's surface consists of water (oceans or lakes), the radiation can penetrate and be absorbed through a considerable thickness, so that the same amount of heat is spread through a larger mass, resulting in a much smaller temperature increase. Other factors also contribute to reducing the temperature rise when radiation is absorbed by a water surface: evaporation uses some of the heat, and the motions of the water may spread it through even deeper layers than the radiation penetrates.

The ground heated by radiation tends to heat the air above it. It does so in two ways: by contact, conduction, and convection—the way in which the burner of a stove heats a pot of water; and by radiation. The ground radiates approximately as a black body at its own temperature. This radiation passes upward through the overlying atmosphere, and some of it is absorbed by the layers of air it passes through and some escapes into space. As shown in Figure 3.3B, at the temperatures that occur at the ground the emitted radiation is in the far infrared, with maximum energy at about 10 $\mu$m. [Applying equation (3.4) to the average temperature of the earth's surface we have $\lambda_M = 2898/288 = 10.1$ $\mu$m.] At many wavelengths in this region, water vapor and carbon dioxide are good absorbers, and the air layers near the ground absorb a large part of the radiation. It is this heating from below by contact and by radiation that is responsible for the warmth of the lower layers of the atmosphere and the decrease of temperature upward through the troposphere.

Figure 3.11 is a graph of the total radiation emitted from a black body for temperatures that occur in the earth's atmosphere. The energy emitted by a black body at the mean temperature of the earth's surface is seen to be approximately 390 W/m$^2$.

In the next section we shall examine in more detail the exchange of the long-wave (infrared) radiation.

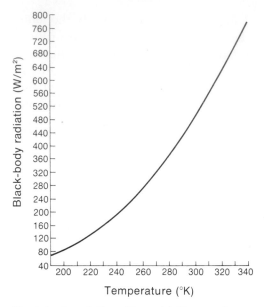

**Figure 3.11** *Black-body radiation at terrestrial temperatures.*

## 3.7 Transfer of terrestrial radiation through the atmosphere

The radiation from the earth's surface is principally in the range of wavelengths from 4 to 70 $\mu$m, with peak intensity in the vicinity of 10 $\mu$m, whereas the radiation from the sun that reaches the earth consists almost entirely of wavelengths shorter than 4 $\mu$m. Thus there is practically no overlap. Because of this, the radiation from the sun is frequently called *short-wave radiation*, and that from the earth *long-wave radiation*. Since the atmosphere is also at relatively low temperatures compared with the sun, the radiation from it is also long-wave radiation. Consisting of gases, however, the atmosphere does not radiate in all wavelengths, like a black body, but only in certain wavelengths, that is, the wavelengths at which it can absorb.

Figure 3.12 shows the relative amount of radiation that can be absorbed by some of the gases in the atmosphere and by the atmosphere as a whole. It shows that the atmosphere is much more transparent for short-wave than for long-wave radiation, particularly in the visible. For long-wave radiation there is a band between 8 and 11 $\mu$m in which the atmosphere absorbs very little. The absorption that shows in the middle of this band, centered at 9.6 $\mu$m, is due to ozone in the stratosphere, so that the troposphere is practically completely transparent in this band.

The result of the atmosphere being transparent to solar radiation and more absorptive to long-wave radiation from the earth is that the earth's surface is kept at a higher temperature on the average than it would have if there were no atmosphere. The radiant energy absorbed by the atmosphere is partially radiated back to the earth's surface, increasing the total

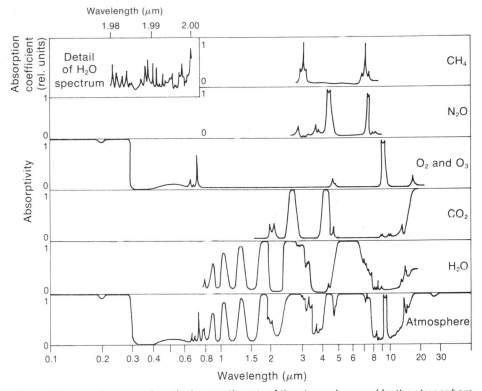

**Figure 3.12**    *Absorptivity at various wavelengths by constituents of the atmosphere and by the atmosphere as a whole. [From R. G. Fleagle and J. A. Businger,* An Introduction to Atmospheric Physics. *Copyright 1963 by Academic Press.]*

energy received there. The raising of the surface temperature because of the back-radiation from the atmosphere is known as the *greenhouse effect*, because it was thought that the glass roof of a greenhouse, being transparent to solar radiation and relatively opaque to long-wave radiation, plays the same role as the atmosphere in keeping the temperature in the greenhouse high enough for plants to grow in winter. However, it has been shown that in greenhouses the prevention of air movement is the more important factor; in small greenhouses in which the glass panes in the roofs were replaced by rock salt, which allows long-wave radiation to escape, the temperature was about as high. For this reason it has been suggested that the increase in temperature due to trapping of radiation by the air be called the atmospheric effect instead of the greenhouse effect. However, we shall use the customary term, while recognizing that it is based on a misconception.

Figure 3.12 shows that the principal absorbers of long-wave radiation are water vapor and carbon dioxide. Water vapor absorbs strongly at wavelengths between 5 and 7 $\mu$m and longer than 12 $\mu$m, and the carbon dioxide between 4 and 5 $\mu$m and longer than 14 $\mu$m. The band between 8 and 11 $\mu$m, which is practically transparent except for the absorption at 9.6 $\mu$m by stratospheric ozone, is called the *atmospheric window*. Radiation from the ground or cloud tops in these wavelengths passes directly through the troposphere, and all but the portion absorbed at 9.6 $\mu$m goes out to space unimpeded. In the other wavelengths, radiation from the ground is absorbed at various levels in the atmosphere, and in turn the atmosphere at those levels radiates up and down in amounts that depend on its temperature and water vapor content.

Figure 3.13 illustrates schematically the way in which the various layers of air gain or lose energy by radiative exchange. The troposphere is assumed to be divided into three layers of equal absorptivity due to water vapor and carbon dioxide. The double arrows pointing into the layers and into the ground represent the amount of radiation absorbed there and not, as in Figure 3.10, the amount passing through. The single arrows represent radiation emitted by the layers or by the ground, or the amount of radiation passing through the tropopause level at the top of layer 3.

It will be convenient to use a general designation for the various quantities pertaining to the layers, so that, for instance, when we write $T_i$ ($i = 1, 2,$ or 3) we mean the temperature of any one of the layers. Each layer emits radiation $E_i$ both up and down in an amount that depends on its temperature and its water vapor and carbon dioxide content. Since we have assumed equal amounts of these gases, the emissions of the layer will depend on the temperature only. The figure is drawn assuming a normal decrease of temperature with height ($T_3 < T_2 < T_1$), so that the amount of radiation emitted, indicated by the length of the arrows, is correspondingly smaller for the higher layers, that is, $E_3 < E_2 < E_1$. Note that the total amount of energy emitted by each layer is $2E_i$.

The emission from the ground, designated by $E_s$, is approximately black body radiation. Since it includes radiation in wavelengths (between 8 and 11 $\mu$m) in which the air layers do not emit, $E_s$ will be considerably greater than $E_1$ for this reason, in addition to $T_s$ being greater, during the day at least, than $T_1$. Part of the radiation from the ground will be absorbed by layer 1. Since much of $E_s$ in the wavelengths absorbed by water vapor and carbon dioxide will be removed by the time it has passed through layer 1, a smaller part will be absorbed by layer 2, and, similarly, the amount absorbed by layer 3 will be still smaller. If we designate the part of $E_s$ that is absorbed by the $i$th layer $A_{si}$, we have seen that $A_{s1} > A_{s2} > A_{s3}$. But the part of $E_s$ that is between 8 and 11 $\mu$m will pass through all the layers practically unabsorbed; thus an amount $E_{sT}$ is emitted through the tropopause from the ground.

The emission up or down from each layer may similarly be traced as shown in Figure 3.13. For instance, of the emission upward from layer 2, a part $A_{23}$ is absorbed by layer 3 and the rest, $E_{2T}$, passes through the tropopause, while of its downward emission layer 1 absorbs $A_{21}$ and the ground receives $A_{2s}$.

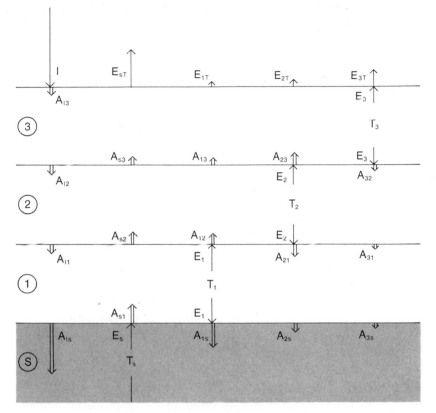

**Figure 3.13**  *Illustration of radiative transfer through the atmosphere: single arrows — radiation emitted by ith layer, $E_i$ (or solar radiation I); double arrows — radiation $A_{ij}$ absorbed by jth layer from $E_i$. [See text for explanation.]*

We may similarly treat the solar radiation passing downward through the layers. Of the amount $I$ passing through the tropopause, $A_{I3}$ is absorbed by layer 3, $A_{I2}$ by layer 2, and $A_{I1}$ by layer 1. Because of the small absorptivity of air for short wavelengths (once the radiation at wavelengths shorter than 0.3 $\mu$m has been removed), these quantities are small; the remainder, $A_{Is}$, is the amount absorbed at the ground. Of course $I$ varies with time of day — maximum at noon and zero at night — and $A_{Is}$ undergoes a similar variation. At any one time the absorption of radiation at the ground is

$$A_s = A_{Is} + A_{1s} + A_{2s} + A_{3s}$$

The net amount of energy available for heating the ground is

$$H_s = A_s - E_s$$

During the day, when the sun is high, $A_s$ is larger then $E_s$, and the temperature at the ground rises. During the night, when $A_{Is} = 0$, $A_s$ is smaller than $E_s$, and the surface temperature decreases.

We can likewise discuss the radiation budget for each of the layers. For instance, the net amount of energy added to layer 2 by radiation is

$$H_2 = A_{l2} + A_{s2} + A_{12} + A_{32} - 2E_2$$

It turns out that usually the layers of air away from immediate contact with the ground are nearly in radiative equilibrium, that is, $H_i$ is very small in these layers. During the day there may be a slight surplus of radiation, but on the average through the day and night there is a net deficiency that must be made up for by other processes. The processes by which the deficit in the troposphere is made up are conduction, convection, and transfer of latent heat by evaporation from the earth's surface and condensation to form clouds.

In the long run these processes result in a balance between the incoming and outgoing energy for the earth and atmosphere and for each separately. Figure 3.14 shows the estimates of the component streams of radiation and the other processes that produce this balance. The radiation from the sun on the earth, averaged for the entire year, is represented by 100 units, of which less than half is absorbed at the ground. Due to the greenhouse effect, which produces a higher surface temperature than would occur in the absence of the atmosphere, the earth's surface emits 114.5 units, most of which is absorbed by the atmosphere. The atmosphere in turn radiates back to the ground 96.5 units, more than twice that absorbed at the ground from the sun, providing the excess that heats the atmosphere from below by conduction, convection, and transfer of latent heat and thus contributing to the decrease in temperature with height through the troposphere.

We have seen that the high atmosphere, above 100 km, is heated, dissociated, and ionized by the very short ultraviolet radiation that is strongly absorbed by atomic and molecular oxygen, that the layer around 80 km is shielded from the radiation it could absorb by the layers above it, and that below this, where the air gets dense enough for ozone to form by collisions, there is a layer that is heated by the absorption of ultraviolet radiation by ozone. The ultraviolet radiation at wavelengths below 0.3 $\mu$m is completely absorbed by the ozone layer. The longer-wave radiation passes through this layer and is absorbed only slightly in the lower stratosphere and the troposphere on its way to the ground. The warmth of the lower part of the troposphere is due to heating from below by long-wave radiational exchange, transfer of latent heat, conduction, and convection.

Quantitative studies have shown that these processes explain the observed average temperature structure in the vertical. Similarly, computations of the ionizing effects of the radiation result in ion density distributions corresponding to that which is observed. These computations demonstrate that the way in which solar radiation is absorbed and transmitted through the atmosphere provides the explanation of the remarkable variation of temperature and composition with height.

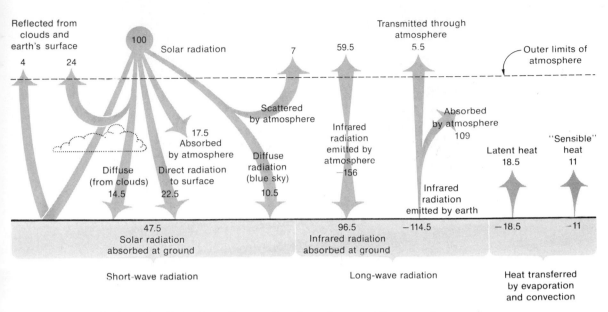

**Figure 3.14**  *Budget of radiation from the sun, the atmosphere, and the ground.*

To understand in detail the variations of temperature with time and with latitude we must examine how the air behaves when it is heated or cooled. This we shall do in the next chapter.

Questions, Problems, and Projects for Chapter 3

1  Define the following terms:

| | | |
|---|---|---|
| a  Radiation | e  Wave speed | i  Solar constant |
| b  Wave | f  Energy | j  Scattering |
| c  Wavelength | g  Calorie | k  Absorption |
| d  Frequency | h  Black body | l  Reflection |

2  The earth is continually receiving radiation from the sun. Why doesn't its temperature continually get higher and higher?

3  Discuss the reason why the time from sunrise to sunset varies with time of year and with latitude. Why is it always 12 hours at the equator?

4  What would be the consequences of removing ozone ($O_3$) from the upper atmosphere? What implications would this have for the flight of supersonic aircraft, which are designed to fly at about 22 km?

5  How does the amount and dominant wavelength of radiant energy from a black body vary with temperature? Discuss the radiation from the sun and the earth's surface in these terms.

6  Why is the diurnal variation of temperature over land larger than over the sea?

7  Discuss what happens to solar radiation in passing through the atmosphere, from the "top" of the atmosphere to the ground.

8  What constituents of the atmosphere absorb long-wave radiation? How would changes in the amount of carbon dioxide affect the temperature near the earth's surface?

# FOUR
## The Gas
## Laws.
## Heat and
## Temperature
## Changes

### 4.1  The behavior of gases

The gas phase of a substance differs from the other two phases, liquid and solid, in that a gas tends to occupy all the space available to it. If a small amount of gas is placed in a closed container it does not form a free surface, as a liquid does, but spreads throughout the entire volume. The molecules of a gas are not bound to one another, but move about freely, except for bouncing against each other or the walls of the container. When one molecule collides with another or with the wall, it rebounds like an elastic ball and in so doing exerts *pressure*. Until the molecules are uniformly spread through the container the pressure will be uneven and the molecules will be pushed more toward the places where there are fewer of them pushing back. Very quickly an equilibrium state is reached, in which the internal pressure of the gas either is constant or varies in such a way that it just balances any external forces, such as gravity, that are acting on the gas.

Figure 4.1 illustrates another kind of experiment we might carry out. In each of the diagrams the situation is illustrated in which a particular quantity of gas has been placed in a container that has a movable top, that is, a cylinder with a piston at the top. (For the purpose of this illustration we assume the space above the piston to contain no matter.) If the piston were weightless, the pressure on it due to the bombardment of it by the gas

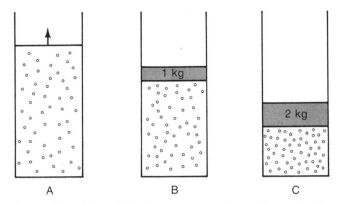

**Figure 4.1** *Illustration of the effect of pressure on volume of gas when the temperature is kept constant: (A) weightless piston moves upward because of pressure of gas; (B and C) piston weighing 2 kg compresses gas to one-half the volume and twice the density of that compressed by a piston weighing 1 kg at the same temperature.*

molecules would push it upward indefinitely, as indicated in diagram A. Since a real piston has weight, it will be pushed to the position where the upward force due to the gas pressure inside the cylinder is just equal to the weight, as shown in diagram B. If we increase the weight, the piston will be pulled down by the force of gravity, squeezing the molecules of gas together and increasing the number colliding with it until the increase in pressure force so produced is just equal to the increase in weight. This downward displacement of the piston corresponds to a decrease in volume and an increase in density of the gas. If the weight on the piston is doubled and at the same time the temperature is kept constant (diagram C), the volume will be halved and the density will be doubled.

The average speed of the molecules, and therefore the force they exert when they collide, varies with the temperature. If the temperature is raised in a rigid container (constant volume), the pressure will rise because of the increased speed of the molecules. If the temperature is raised in a cylinder with a piston of fixed weight (constant pressure), the piston will be pushed upward and thus the volume increased and density decreased because of the increased speed of the gas molecules.

The behavior of gases can thus be expressed in the following rules:

1 The density of a gas is proportional to the pressure, and the volume is inversely proportional to the pressure, if the temperature is constant.

2 The pressure of a gas is proportional to its absolute temperature if the volume is constant.

3 The volume of a gas is proportional to the absolute temperature, and the density is inversely proportional to it, if the pressure is constant.

**Figure 4.2** *Pressure measuring instruments. A. Mercury barometer, in which the atmospheric pressure is balanced by the mercury in an evacuated tube. The length of the mercury in the tube is a measure of the pressure, and, consequently, pressure is frequently expressed in terms of length. Thus standard atmospheric pressure (1013.25 mb) is 760 mm of mercury. B. Aneroid barometer, in which the atmospheric pressure compresses a partially evacuated flexible metal box. The changes in the size of the box with pressure are converted by a mechanical linkage to the position of the pointer on the dial. In the example shown, the scale shows the pressure in millibars (inner scale) and inches of mercury (outer scale). [Courtesy of Science Associates, Inc.]*

These rules can be expressed quantitatively in a single equation, which is called *the equation of state of a perfect gas,* or *the gas law*:

$$p = \rho RT \tag{4.1}$$

or

$$p\alpha = RT$$

where the letters have the following meanings:

$p$   Pressure

$\rho$   Density

$T$   Absolute temperature

$R$   A constant of proportionality having a particular value for each gas in any particular system of units. $R$ is called the *gas constant*. In the cgs system its value for dry air is $2.87 \cdot 10^6$ ergs/gram °K; in the mks system, 287 joules/kilogram °K.

$\alpha$   Specific volume

We have used the terms pressure, density, and temperature without defining them, and while they have meanings for us gained from common experience, it is desirable to make these meanings precise:

*Pressure* is the force per unit area acting perpendicular to the area. On the walls of the container it is the force pushing outward on the wall because of the bombardment by the molecules of gas. In the interior of the gas it is the force that the molecules on one side of an (imaginary) unit area exert on those on the other side, and is equal and opposite to the force that is exerted by those on the other side. Since we can have such imaginary unit areas constructed in all directions through a point, *the pressure at a point is the same in all directions.*

The unit of force in the cgs system is the dyne, where one dyne is the force that, acting on a mass of one gram, would produce an acceleration of one centimeter per second per second, and the unit of pressure is the dyne per square centimeter. Similarly, in the mks system the unit of force is the newton, defined as the force that would accelerate a mass of one kilogram one meter per second per second, and the unit of pressure is the newton per square meter. These pressure units are so small that normal or standard pressure at sea level is 1,013,250 dynes/cm² or 101,325 newtons/m². To avoid such big numbers in meteorological practice, pressures are commonly expressed in *millibars* (mb), where 1 mb = $10^3$ dynes/cm² = $10^2$ newtons/m². Normal sea-level pressure is 1013.25 mb.

*Temperature* is the property of a body[1] that determines whether heat will flow to or from it when it is placed in contact with another body. If there is no flow of heat between two bodies, they are at the same temperature; otherwise the body that receives heat from the other is at the lower temperature. Absolute temperature can be measured by the expansion of a gas at constant pressure or the change in its pressure in a constant volume. More conveniently, for most purposes temperature is measured by the expansion of a liquid in a glass container (relative to the expansion of the container).

The Celsius (centigrade) scale of temperature has 0° for the melting point of ice and 100° for the boiling point of water at normal sea-level atmospheric pressure. It was named for Anders Celsius, a Swedish astronomer, who in 1742 introduced a scale with 100 degrees between ice point and boiling point, but with the 0 and 100 reversed. The Celsius scale has

---

[1] By "body" we mean any delimited or circumscribed quantity of matter. Thus we might refer to a book as a body, or the gas in a particular volume as a body.

come into general use in science, and, except in English speaking countries, it is used in everyday affairs as well. England is in the process of going over to the Celsius scale; weather reports are broadcast there with temperatures in Celsius, with the Fahrenheit equivalents given parenthetically.

Gabriel Daniel Fahrenheit, who was born in Germany, spent most of his life in Holland as a manufacturer of and experimenter with meteorological instruments. He introduced the mercury thermometer and his temperature scale in 1714. The first thermometer, introduced a century earlier, used the expansion of air and was subject to many sources of error. Subsequently, sealed tubes containing alcohol were used. Individually, these gave good readings, but the readings from one to another could not be compared for lack of a reference scale. Fahrenheit showed that the freezing point of pure water in the presence of pure ice is always the same, and he invented the mercury thermometer to test whether the boiling point of water is also constant. For zero on his scale of temperature he chose the lowest temperature he could attain by a mixture of ice and common salt, and for the other fixed point on his scale he used body temperature, which he arbitrarily called 96°. On this scale he found the freezing point of water to be 32°, and the boiling point 212°, and these became the standard reference points for the Fahrenheit scale. With improved thermometers the average human blood temperature was shown to be 98.6°F instead of 96°F.

The absolute scale is called the Kelvin scale after William Thomson, Lord Kelvin, who in 1848 showed the thermodynamic significance of absolute temperature. Its zero is based on equation (4.1) and the size of one degree, the Celsius degree, which is one one-hundredth of the temperature change from the freezing point to the boiling point. Experiments have determined that $0°K = -273.15°C$, or $0°C = 273.15°K$.

In meteorology in the United States all three temperature scales are used, and it is important to be familiar with the equivalents in the different scales. The conversion relationships are given in Appendix A.

It facilitates computations to note that $1.8 = 2 - 0.2$ and $5/9 = 1/2 + 1/20 + 1/200 + \ldots$ . For example, to convert 20°C to °F, we simply multiply by two, giving 40°, subtract one-tenth of that, or 4°, giving 36°, and add 32°, giving 68°F. To convert $-28°F$ to °C, we first subtract 32, giving $-60°$, take one-half, $-30°$, and add one-tenth of that, $-3$, one one-hundreth, $-0.3$, and so forth, obtaining $-33.333 \ldots °C$. For approximate conversions the scales in Figure 4.3 may be used.

*Density* is the mass of unit volume of a substance. Mass is the measure of the amount of matter; it is the property that provides inertia, that is, resists changes in amount of motion, and provides gravitational attraction of other bodies. From the latter characteristic we have, for bodies on earth,

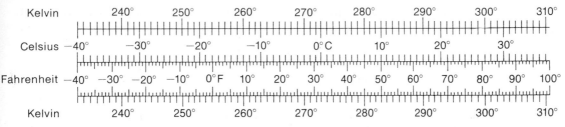

**Figure 4.3** *Relationship between Fahrenheit, Celsius (centigrade), and Kelvin (absolute) temperature scales.*

the convenient method of measuring the mass of a body by weighing it. If the body's mass is $M$, its weight is $Mg$, where $g$ is the attracting force that the earth exerts on unit mass. The quantity $g$ is called the *acceleration of gravity*. If the density varies from point to point we must define it as the limit of the ratio of the mass of a very small volume of the substance to the volume as the volume gets smaller and smaller.

The unit of mass in the cgs system is the gram, originally defined as the mass of one cubic centimeter of water, and the unit of density is the gram per cubic centimeter. Thus the density of water in cgs units is approximately 1.0 g/cm³. We can compute the density of dry air by using equation (4.1). Solved for the density, we have

$$\rho = p/RT$$

As an example we shall do so for "NTP" (normal temperature and pressure), which is a temperature of 0°C and standard sea-level pressure, 1013.25 mb. In cgs units,

$$p = 1.01325 \cdot 10^6 \text{ dynes/cm}^2$$

and

$$R = 2.87 \cdot 10^6 \text{ ergs/gram °K}$$

Converting 0°C to Kelvin, we have

$$T = 273.15$$

Substituting these values in the equation we have

$$\rho = \frac{1.01325 \cdot 10^6}{2.87 \cdot 10^6 \cdot 273.15} = 0.001293 \text{ g/cm}^3 = 1.293 \cdot 10^{-3} \text{ g/cm}^3$$

This value is about 1/800 the density of water.

In the mks system the unit of mass is the kilogram, and the unit of density is the kilogram per cubic meter. Density in mks units is 1,000 times that in cgs units. Water has a density of 1,000 kg/m³, air at standard conditions, 1.293 kg/m³.

*Specific volume* is the volume occupied by unit mass of a substance. It is thus the reciprocal of the density, that is,

$$\alpha = 1/\rho \tag{4.2}$$

In the cgs system its unit is cubic centimeters per gram; in the mks system cubic meters per kilogram. At standard conditions air has a specific volume of 775 cm³/g or 0.775 m³/kg.

When using any physical equation, such as the equation of state, all quantities must be expressed in the same system of units, for example, either cgs units or mks units.

## 4.2 Variation of pressure with height

We are now in a position to explain why the pressure (and the density) decreases with height in the atmosphere. Consider an air column of *unit* cross section from the ground up to the "top" of the atmosphere (Figure 4.4). The pressure at any arbitrary height $h$ in the air column is equal to the weight of the part of the air column above it, in the same way that the pressure inside the cylinder considered in the previous section is equal to the weight of the piston.

Suppose that the weight of the air column above height $h$ is $W$. If we designate the pressure at that height $p_A$, we have $p_A = W$. Now consider a position $B$ a distance $\Delta h$ above $A$.[2] The weight of the air column above $B$ will be less than $W$ by the weight of the part of the air column between $A$ and $B$. Since the column has unit cross sectional area, the volume of this portion is equal to $\Delta h$, and its weight is therefore $\rho\, g\, \Delta h$. The pressure at $B$ is therefore

$$p_B = W - \rho\, g\, \Delta h = p_A - \rho\, g\, \Delta h$$

The change in pressure in going up a distance $\Delta h$ is thus

$$\Delta p = p_B - p_A = -\rho\, g\, \Delta h \tag{4.3}$$

This equation is called the hydrostatic equation.

If we now substitute the value of $\rho$ from the gas law, equation (4.1), we have

$$\Delta p = -\frac{pg}{RT}\, \Delta h \tag{4.4}$$

From equation (4.3) we see that if we go up a distance $\Delta h$ we will experience a decrease in pressure that is proportional to the pressure and inversely proportional to the temperature. Near sea level, where the

---

[2]The notation $\Delta h$ means that we are considering a small increase or increment in the quantity $h$. The symbol $\Delta$ (Greek letter *delta*) before a quantity means a small increment in it. It is important to distinguish this notation, in which $\Delta h$ is a single quantity, from the usual algebraic notation, in which $ab$ means the product of quantity $a$ times quantity $b$.

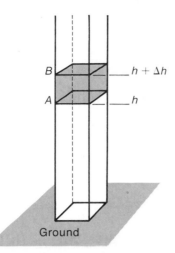

**Figure 4.4** *Diagram illustrating the variation of pressure with height. Difference in pressure between A and B is the weight of the air column of unit cross section between the two levels.*

pressure is high, the decrease in pressure for a given increase in height is large, and at great heights, where the pressure is small, the rate of decrease in pressure with height is much smaller.

The temperature also appears in equation (4.4). However, temperature changes much less rapidly with height than pressure, and to a first approximation its variation can be ignored. The change of pressure for a given change of height is thus approximately a constant fraction times the pressure itself. A quantity that changes a fixed fraction of itself for each change of a fixed amount in another variable is said to vary *exponentially*, and thus pressure decreases approximately exponentially with height. It is reduced to one-half its sea-level value at a height of about 5.5 km, one-fourth at about 10 km, one-eighth at about 15 km, and one-sixteenth at about 20 km.

As an example of the application of equation (4.4), let us compute the approximate change of pressure in going upward 10 m if the pressure at sea level has its standard value, 1013.25 mb, and the temperature is 15°C. In mks units and °K we have

$$p = 1.01325 \cdot 10^5 \text{ newtons/m}^2$$
$$R = 287 \text{ J/kg °K}$$
$$T = 288.15°K$$
$$g = 9.8 \text{ m/sec}^2$$
$$\Delta h = 10 \text{ m}$$

$$\Delta p = -\frac{1.01325 \cdot 10^5 \cdot 9.8 \cdot 10}{287 \cdot 288.15} = -1.20 \cdot 10^2 \text{ newtons/m}^2 = -1.2 \text{ mb}$$

**Table 4.1** U.S. Standard Atmosphere to the Stratopause

| Height (km) | Temperature (°K) | Pressure (mb) | Height (km) | Temperature (°K) | Pressure (mb) |
|---|---|---|---|---|---|
| 0 | 288.2 | 1013.2 | 11.0 | 216.8 | 227.0 |
| 1.0 | 281.8 | 898.8 | 12.0 | 216.6 | 194.0 |
| 2.0 | 275.2 | 795.0 | 14.0 | 216.6 | 141.7 |
| 3.0 | 268.7 | 701.2 | 16.0 | 216.6 | 103.5 |
| 4.0 | 262.2 | 616.6 | 18.0 | 216.6 | 75.65 |
| 5.0 | 255.7 | 540.4 | 20.0 | 216.6 | 55.29 |
| 6.0 | 249.2 | 472.2 | 25.0 | 221.6 | 25.49 |
| 7.0 | 242.7 | 411.1 | 30.0 | 226.5 | 11.97 |
| 8.0 | 236.2 | 356.5 | 35.0 | 236.5 | 5.75 |
| 9.0 | 229.7 | 308.0 | 40.0 | 250.4 | 2.87 |
| 10.0 | 223.2 | 265.0 | 50.0 | 270.6 | 0.78 |

From this we see that near sea level, for typical terrestrial temperatures, the pressure decreases approximately one millibar for every ten meters we go upward.

The U.S. Standard Atmosphere is defined in terms of temperatures and height, as given in Table 2.3. The corresponding pressures at various heights up to the stratopause are given in Table 4.1.

Equation (4.4) is also used in evaluating elevations determined by altimeters and in reducing the pressure observed at weather stations to sea level. The usual altimeter consists of an aneroid barometer with the pressure scale replaced by a height scale that gives the elevation in the standard atmosphere at which the pressure occurs. In most altimeters the zero of the altitude scale can be shifted to allow for the departure of the sea-level pressure from its standard value, and for an additional correction to compensate for the departure of the actual temperature from that in the standard atmosphere. For aircraft, radio altimeters are available that determine the height above the terrain by measuring the time it takes for a radio signal sent from the plane to be reflected back from the ground below.

To enable comparison, the pressure measured at various weather stations, which are at different elevations, is "reduced to sea level." In effect, the pressure difference $\Delta p$ of an imaginary air column with thickness $\Delta h$ equal to the altitude of the station and an appropriate average temperature $T$ is added to the station pressure. The temperature used is an estimate of the average temperature an air column from the station level to sea level would have if the solid earth between were absent, based on the observed temperature at the station.

## 4.3 Heat and temperature change

Heat is a form of energy. It is associated with the random motions of the molecules of which matter is composed. In solids and liquids these motions consist largely of vibrations of the molecules; in gases they are the free movements of the molecules between collisions. The temperature is a measure of the kinetic energy of the molecules. If heat is added to a quantity of matter, the speed of the molecules, and thus the temperature, will be increased. This can be expressed by the equation

$$H = cM\ (T_2 - T_1)$$

where $H$ is the amount of heat added, $M$ is the mass (amount of matter), $T_1$ and $T_2$ are the temperatures before and after the heat was added, and $c$ is a constant of proportionality that is called the *specific heat of the substance.*

If we are dealing with a gas, we must distinguish the circumstances under which the heating takes place. If the gas is in a closed container, so that the volume is constant, a given amount of heat will produce a larger temperature change than if the gas is in a container with a movable piston, which rises so that the pressure is kept constant. This is because in the latter case some of the energy must be used to do the work of raising the piston. Therefore gases are considered to have two specific heats, $c_v$ and $c_p$, where $c_v$, the specific heat at constant volume, is less than $c_p$, the specific heat at constant pressure.

We shall be dealing with phenomena in the open air where, on the one hand, the air is free to expand, so that the volume of a given quantity (say, unit mass) of air will not remain constant, but at the same time the air may rise or fall, so that the pressure will change also. For convenience we shall consider what happens to unit mass (1 g or 1 kg) of air. It can be shown that the heat required per unit mass to produce a change from $T_1$ to $T_2$, when both the volume and the pressure change, may be written in terms of either $c_v$ or $c_p$ as follows:

$$H = c_v\ (T_2 - T_1) + p\ (\alpha_2 - \alpha_1)$$
$$H = c_p\ (T_2 - T_1) - \alpha\ (p_2 - p_1)$$

These equations are strictly true only when $H$ is very small, so that $T_2 - T_1$, $p_2 - p_1$, and $\alpha_2 - \alpha_1$ are also very small. To indicate this we let

$$T_2 - T_1 = \Delta T$$
$$p_2 - p_1 = \Delta p$$
$$\alpha_2 - \alpha_1 = \Delta \alpha$$

and write $\Delta H$ instead of $H$, where, as explained in footnote 2 (Section 4.2), $\Delta$ is understood to represent a small change in the quantity represented by the letters that follow. The equations then become

$$\Delta H = c_v\ \Delta T + p\ \Delta \alpha \qquad (4.5)$$
$$\Delta H = c_p\ \Delta T - \alpha\ \Delta p \qquad (4.6)$$

Except in the layers close to the ground and in the high atmosphere, heat is added to (or subtracted from) the air very slowly and the air is usually moving fast, so that if we are considering changes over a short time (say, less than one day) in the upper part of the troposphere and lower stratosphere the exchange of heat between an air parcel and its environment can be neglected. A process in which no heat exchange between an air parcel and its surroundings occurs is called *adiabatic,* and thus motions of the air are approximately adiabatic, except near the ground.

For processes in which the pressure changes adiabatically we set $\Delta H = 0$ in equation (4.6). Solving for $\Delta T$, we get

$$\Delta T = \frac{\alpha}{c_p} \Delta p \tag{4.7}$$

If we substitute for $\Delta p$ from equation (4.4) we get

$$\Delta T = -\frac{g}{c_p} \Delta h$$

or

$$\frac{\Delta T}{\Delta h} = -\frac{g}{c_p} \tag{4.8}$$

Since $g$, the acceleration of gravity, and $c_p$, the specific heat at constant pressure, are constant we see that the change in temperature for a given change in height in an adiabatic process is a constant.

We represent this adiabatic rate of temperature change by $\Gamma$ (Greek capital letter *gamma*)

$$\Gamma = \frac{g}{c_p} \tag{4.9}$$

The value of $\Gamma$ depends on the choice of units for $\Delta h$. It is convenient to remember its value for $\Delta h$ expressed in hundreds of meters:

$$\Gamma = 0.98°C/100 \text{ m}$$

For practical purposes the value can be taken as one degree per hundred meters. This rate applies only to cloudless air; when clouds form or are present other processes affect the rate of temperature change. We may therefore state the following rule:

*If unsaturated air rises (falls) without receiving or losing heat it will cool (warm) one Celsius degree for each 100-meter change in height.*

This suggests part of the reason why the temperature decreases in the troposphere, for in moving over the rough and unevenly heated surface of the earth it is bound to rise and fall. The displaced air will mix with the surroundings at its new position, lowering the temperature aloft and raising it at low levels.

## 4.4    Thermodynamic diagrams. Process curves and sounding curves

Let us consider a small mass of air (which we refer to as an *air parcel*). As the parcel moves (or even when it remains stationary) its pressure and temperature will vary. For instance, it may lose heat by emitting radiation faster than it regains it by absorbing radiation, without the pressure changing; in this case we say it undergoes cooling during an *isobaric process.* Or it may move horizontally across isobars or up or down, with associated changes in pressure. If it does not receive or lose any heat while moving, its temperature will change at the adiabatic process rate, or if heat is added or subtracted during its motion its temperature will undergo some other variation. To study these processes it is convenient to represent the changes in the state of an air parcel graphically. A diagram on which variations of the thermal state of a system are shown is called a *thermodynamic diagram.*

Thermodynamic diagrams may have any two of the quantities in the equation of state, equation (4.1), as coordinates. For instance, in one diagram that is used when studying heat engines in physics or engineering the pressure is plotted against volume (the cylinder volume or the specific volume of the gas it contains). This choice of coordinates is made because on the pressure-volume (P–V) diagram, work done in a cyclic process is proportional to the area enclosed by the curve representing the process. In meteorology it is convenient to use the directly measured quantities, pressure and temperature, as coordinates. By choosing appropriate scales for the coordinates, the work-area property of the P-V diagram can be retained. One such combination of scales is the pressure on a logarithmic scale and the temperature on a linear scale.

Figure 4.5 shows a thermodynamic diagram with temperature ($T$) as the abscissa, evenly spaced, and the logarithm of the pressure (really, $\log p_0/p$, where $p_0$ is a constant, for instance, 1000 mb) as ordinate. Note that the logarithmic scale results in larger spacing for the same pressure difference as the pressure decreases. A consequence of this choice is that the ordinate corresponds to height; distances upward on the diagram are approximately proportional to distances upward in the atmosphere.

The state of the air parcel (its temperature and pressure) is represented by a point on this diagram. Isobaric processes are represented by horizontal lines (isobars). Isothermal processes in which pressure varies but heat is added (or subtracted) at exactly the right rate to keep the temperature constant would be represented by vertical lines (isotherms). Lines representing adiabatic processes are called *adiabats,* in this case *dry adiabats,* since we are dealing with cloudless air. Dry adiabats for two examples of adiabatic processes are drawn on the diagram. Line *AB* represents the variation of temperature for a parcel of air having an initial temperature of 0°C as its

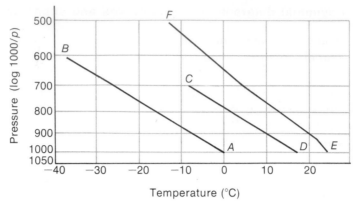

**Figure 4.5** *Thermodynamic diagram: AB and CD are process curves representing the change in temperature with adiabatic change in pressure;
EF is the sounding curve representing the variation of temperature
with pressure in the air above Washington, D.C., 7:00 P.M., July 1,
1957.*

pressure changes adiabatically from 1000 mb to 600 mb. At 600 mb its temperature would be −37.3°C.

The temperature a parcel of air would attain if it were to undergo an
adiabatic compression or expansion until the pressure reaches 1000 mb is
called the *potential temperature*. Line *CD* in Figure 4.5 represents the
adiabatic process a parcel having an initial temperature of −10°C and
pressure of 700 mb would follow in being compressed. It is seen that its
potential temperature is 18.5°C, or 291.6°K.

For convenience, adiabatic process curves are printed on the thermodynamic diagrams used routinely in meteorology, in addition to the grid
of coordinate lines. The potential temperature of any parcel whose representative point falls on an adiabat is the same, namely, the temperature at
which the adiabat intersects the 1000-mb isobar. On the diagrams printed
for routine use the adiabats are labeled with the corresponding potential
temperature, in the same way as the isotherms are labeled with the actual
temperature and the isobars with the pressure. Thus the thermodynamic
diagrams used in meteorology have three basic sets of lines printed on
them: isobars, isotherms, and adiabats.

So far we have discussed the representation of processes on a thermodynamic diagram. These processes are changes experienced by an individual air parcel with the passage of time. In addition, curves can be drawn to
represent the state of all the parcels in a vertical air column at a particular
time. This type of curve is called a *sounding curve*.

We have discussed in Chapter 2 the way in which temperature varies
with height in the atmosphere, and in Table 4.1 we presented the variation
of pressure and temperature with height in the Standard Atmosphere. The
Standard Atmosphere represents average conditions at middle latitudes. At
any particular time and place the temperature at various heights (and there-

fore pressures) differs from these standard values. The actual values are measured routinely at hundreds of places all over the earth. The radiosonde measures the temperature and pressure (and the humidity) as the balloon ascends, and sends back the values by radio. The measurement of the properties of the atmosphere in the vertical is called a *sounding;* hence the term sounding curve for the line representing the observed temperatures and pressures along the vertical. Curve *EF* in Figure 4.5 is the sounding curve showing the way in which the temperature and pressure varied over Washington, D.C., at approximately 7:00 P.M. on July 1, 1957.

It is important to keep clear the distinction between process curves and sounding curves. A process curve represents the variation with time of temperature and pressure of a single parcel; a sounding curve represents the measured temperatures and pressures of different parcels of air at various heights in the atmosphere at one particular time.

Since height and pressure are related through the hydrostatic equation [equations (4.3) and (4.4)], process curves and sounding curves can also be represented on a diagram with height and temperature as coordinates. On such a diagram the adiabatic process curves are straight lines.

Usually the sounding curve in a particular instance will not lie along the adiabatic process curve. The variation of temperature with height in an air column thus is generally different from the adiabatic rate of cooling $\Gamma$. The rate of decrease with height of observed temperature as one goes upward in an air column is called the *lapse rate,* usually represented by $\gamma$, the lowercase Greek letter gamma. Thus we may write symbolically

$$-\left(\frac{\Delta T}{\Delta h}\right)_{\text{observed}} = \gamma$$

$$-\left(\frac{\Delta T}{\Delta h}\right)_{\text{adiabatic process}} = \Gamma$$

If the lapse rate in an air column is equal to the rate of adiabatic cooling $\Gamma$, the air column is said to have an *adiabatic lapse rate.*

## 4.5  Atmospheric stability

The ease with which air parcels move up and down relative to the surrounding air depends upon a property called *stability.* The stability of air columns or layers of air is of importance in connection with the diffusion of pollution, the occurrence of turbulence, which may make the flight of airplanes uncomfortable or dangerous, and the development of showers and thunderstorms. To find the criteria for stability we shall consider what happens when one parcel of air in an air column is moved up or down relative to the rest of the column.

We assume that initially all of the parcels in the column are in equilibrium, that is, that the forces acting on them are in balance. (This is the

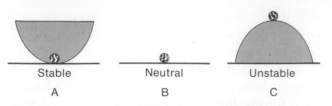

**Figure 4.6** *Illustration of types of stability: (A) marble in stable equilibrium inside bowl; (B) marble in neutral equilibrium on horizontal table top; (C) marble in unstable equilibrium on top of inverted bowl.*

assumption that was made in deriving the hydrostatic equation.) When a parcel is not in equilibrium the unbalanced forces produce an acceleration.

The equilibrium of an object is *stable, neutral,* or *unstable,* depending on what would happen if the object were moved slightly. If the displacement of the object gives rise to forces that tend to bring it back to its original equilibrium position, the equilibrium is said to be *stable.* If the displacement leads to forces that tend to increase the displacement from the equilibrium position, the equilibrium is called *unstable.* If no forces arise from the displacement, the equilibrium is *neutral.*

These types of equilibrium can be illustrated by simple experiments with a marble and a hemispherical bowl. If the bowl is placed on a table with the hollow side up and the marble is placed inside, it will come to equilibrium at the bottom, as in Figure 4.6A. If the marble is pushed a small distance away from this position and released, it will roll back under the force of gravity. (Actually, in falling it will gain momentum that will carry it past the equilibrium position. It will oscillate back and forth until friction damps out its motion.) Thus at the bottom of the bowl the marble is in stable equilibrium. Consider next the situation illustrated in Figure 4.6C, with the bowl inverted and the marble balanced at the top. In principle there is such an equilibrium position, although in practice it may be hard to place the marble exactly at the highest point of the bowl, where it will stay. If the marble is moved the least amount from this position it will continue to roll off the bowl. This situation is thus one of unstable equilibrium. Neutral equilibrium occurs if the marble is at rest on a flat horizontal table (Figure 4.6B). If it is moved a short distance on the table it will remain there; the displacement does not give rise to any force either toward or away from the original position.

The corresponding conditions with respect to the equilibrium of the parcels of air composing an air column are shown in Figure 4.7. In this figure the solid curves represent the sounding curves for four air columns, each of which has a temperature of 14.0°C at a height of 300 m. The column represented by diagram A has a temperature of 13.4°C at 400 m, so that its lapse rate, $\gamma$, is 0.6°C/100 m ($\Delta T = 13.4 - 14.0 = -0.6$°C; $\Delta h = 400 - 300 = 100$ m; $\gamma = -\Delta T/\Delta h = 0.6$°C/100 m). Similarly, for the column represented by diagram B $\gamma = 1.0$°C/100 m, and for the one represented by diagram C

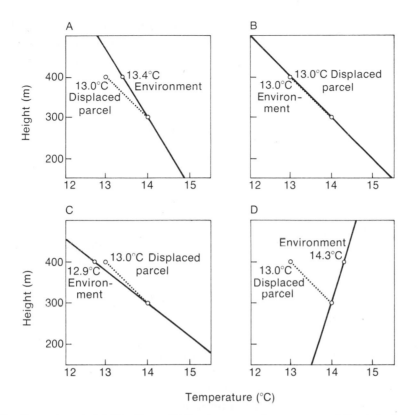

**Figure 4.7**  *Examples of stability. The solid lines are sounding curves and the dotted lines process curves for four schematic cases: (A) normal lapse rate (slightly stable); (B) adiabatic lapse rate (neutral equilibrium); (C) superadiabatic lapse rate (unstable); (D) inversion (very stable).*

$\gamma = 1.1°C/100$ m. Thus the lapse rates of the columns represented by diagrams A, B, and C are, respectively, subadiabatic (normal), adiabatic, and superadiabatic.

Let us consider the temperature of a parcel in each column that is lifted adiabatically from 300 to 400 m in each case. In each instance the parcel would cool from 14.0°C to 13.0°C, the process rate of cooling being shown in the diagrams by the dotted lines. In the case represented by diagram A the displaced parcel would be 0.4°C colder than the surroundings. Being cooler and thus heavier it would tend to fall back to its original equilibrium position. Under these conditions the air column is in stable equilibrium.

In the case represented by diagram B the displaced parcel would have the same temperature as the surroundings, and thus subject to no force, neither one to restore it to its original position nor one tending to move it further away. In its new position it would be in equilibrium, just as it was initially. Under these circumstances the air column is in neutral equilibrium.

In the case represented by diagram C the temperature of the displaced parcel, 13.0°C, is 0.1°C warmer than the undisplaced air column at that level. The displaced parcel of air would thus be lighter than the surrounding air and would experience an upward buoyant force that would tend to displace it farther from its original equilibrium position. The original equilibrium of the air column is thus unstable; any small disturbance of a parcel of it would give rise to forces that increase the departure from the equilibrium position.

It is readily seen that the following criteria apply:

$\gamma < \Gamma$  (subadiabatic lapse rate): stable equilibrium
$\gamma = \Gamma$  (adiabatic lapse rate): neutral equilibrium
$\gamma > \Gamma$  (superadiabatic lapse rate): unstable equilibrium

By examining the consequences of downward adiabatic displacements it will be found that these criteria apply for the stability of the equilibrium of air columns independent of the direction of the displacement.

During the day, when solar radiation is received at the ground and converted to heat there, the heated ground becomes warmer than the air in contact with it, and heat is transferred from the ground to the air. This process leads to the lowest layers becoming unstable, as in diagram C. Any little irregular movement, such as that taking place over uneven ground, causes large vertical displacements that mix the heated air below with the potentially cooler air above. This vertical mixing tends to change the lapse rate from the unstable value towards the adiabatic value. The mixing tends to establish a neutral equilibrium, that is, an adiabatic lapse rate, throughout a layer immediately above the ground in the daytime. The thickness of the adiabatic layer that develops depends on the amount of insolation and the initial stability of the air column. On clear days in summer and at low latitudes the adiabatic layer may extend upward 3 km or more.

The sounding curve in diagram D of Figure 4.7 shows a situation in which the temperature increases with height instead of the normal decrease. A layer of air with this reversal of the normal situation is called an *inversion*. If a parcel were displaced adiabatically up (or down) in an inversion it would be much cooler (or warmer) than the surrounding air and subjected to a very strong restoring force. Because of this, vertical motions are effectively prevented by an inversion. When an inversion is present it suppresses the upward transfer of water vapor or pollutants that are introduced at or near the earth's surface.

At night, with solar radiation absent, the ground loses more energy by its outgoing radiation than it receives from the atmosphere. This is both because it is warmer than the bulk of the water vapor and carbon dioxide that are the principle radiating substances in the atmosphere and because it radiates in all wave lengths, whereas the water vapor and carbon dioxide radiate only in certain wave-length intervals. When the ground cools by radiation the air next to it is cooled by conduction, and later, when the

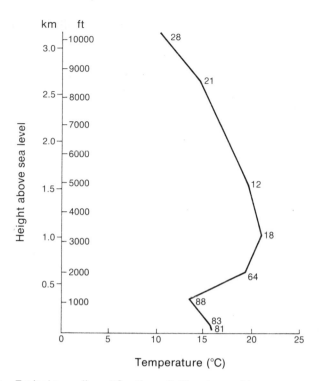

**Figure 4.8** *Typical sounding at Southern California coast in summer (Santa Monica, California, July 1, 1957). Numbers alongside the sounding curve give relative humidity in percent.*

ground is sufficiently colder, by radiation also. The cooling of the air is greatest at the ground and decreases upward. The result is that an inversion is produced next to the ground. This type of inversion, called a *radiation inversion* or *ground inversion,* occurs over land every clear night that the wind is not too strong. In high latitudes in winter, the daytime solar radiation is weak or absent and the ground inversion becomes deeper and more intense day after day.

Another type of inversion is the *trade-wind inversion* or *subtropical inversion.* This type of inversion is present over the eastern part of oceans and the adjoining coasts at subtropical latitudes (equatorward of about 45° latitude) both day and night every day during the warmest one-third of the year and frequently in other seasons. It is produced by the sinking of air as it moves equatorward around the eastern end of the high-pressure areas (anticyclones) that remain over the subtropical oceans most of the time (see Chapter 7). A typical sounding showing this type of inversion is given in Figure 4.8. This sounding shows that at Santa Monica, California (near Los Angeles), on July 1, 1957, there was a 400-m layer with an approximately adiabatic lapse rate, above which an inversion extended to a height of about 1 km. The persistent presence of the inversion at about this level is responsible for the high concentrations of pollution for which Los Angeles has become notorious. Vertical motions carry the pollution emitted by the large urban complex up and down readily through the adiabatic layer, to that extent diluting it. But the inversion above acts as a lid that stops vertical motions. The pollution is kept from mixing with the air at higher levels. Confined to this shallow layer it becomes highly concentrated. Viewed from above the Los Angeles area looks as though it is covered by a lake of brownish smog, the surface of the lake being as sharp and distinct as a water surface. The clear air of the inversion layer is in marked contrast to the murky layer below.

In the schematic diagram of a warm front in Figure 1.5 the warm air is shown flowing upward over the receding cold air. A sounding made in the cold air ahead of the front would pass upward from the cold air into the warm air, and at the frontal boundary surface (really a layer of transition) there would be an increase in temperature with height, that is, a *frontal inversion.* Frequently, the warm air above the frontal surface is only potentially warmer than the cold air below. That is, if it were brought to the same pressure it would be warmer, but at the higher elevation and lower pressure it is actually somewhat colder than the "cold" air that is at a higher pressure. When this situation occurs, the frontal boundary is a layer of greater stability than above or below, but not an actual inversion. A more detailed discussion of fronts and frontal inversions is presented in Chapter 11.

While inversions represent the reverse of what is regarded as the normal temperature variation in the troposphere, they occur frequently and in some places and times persistently. Thus trade-wind inversions are present much of the year not only in Los Angeles, where the severe smog has called worldwide attention to them, but all along the subtropical west coasts of continents and over the adjoining eastern portions of oceans. Like California and Baja California, the coasts of Chile and Peru, and of North Africa and South Africa, are subject to them. Similarly, it has already been mentioned that ground inversions are almost always present over land during clear nights with light winds. Clear nights with light winds are usually associated with high-pressure centers—anticyclones—around which the air spirals outward near the ground. When anticyclones become stationary or move very slowly over land there is sinking motion aloft accompanying the outward spiralling of the air near the surface, which reinforces the surface inversions and may cause them to persist through the day as well as the night. If the stagnation of the anticyclones and the accompanying intense persistent inversions occur over areas where large amounts of contaminants are emitted into the air, high pollution concentrations accumulate. Situations of this sort were responsible for the pollution disasters at Donora, Pennsylvania, in 1948, where 20 people died, and at London, England, in 1952, where there were about 4000 deaths in excess of the normal death rate in one week.

Questions, Problems, and Projects for Chapter 4

1   Define the following terms:

| | | | |
|---|---|---|---|
| a | Temperature | f | Inversion |
| b | Pressure | g | Unstable equilibrium |
| c | Density | h | Lapse rate |
| d | Adiabatic process | i | Isobar |
| e | Sounding | j | Potential temperature |

2   Calculate the Fahrenheit temperatures corresponding to every 10°C from −40°C to +40°C, and make a graph of the relationship on graph paper. Use this graph to make a table of the Celsius temperatures for every 20°F from −40°F to 120°F.

3   If you could make a leakproof balloon of weightless and perfectly elastic material, the pressure inside it would be the same as the pressure of the surroundings, but the density inside it might be different. What could cause the difference? What would happen if the balloon were lifted to a position in the atmosphere where the pressure is lower and allowed to come to thermal equilibrium there?

4   Explain why the air pressure at high levels, for example, mountain tops, is lower than at sea level. Why is it necessary for jet planes to be "pressurized," and why are they equipped with oxygen masks in case of depressurization?

5   When air is pumped into a tire, the tire gets hot. Why? How is this related to the change in temperature of an air parcel when it descends from a higher to a lower level?

6   Consider the variation of temperature with height at various times during a clear day: (a) in the early morning after the air has been cooled by contact with the ground throughout the night; (b) in the midmorning when the ground is being heated rapidly by the sun; (c) at the time of maximum temperature, when it is presumed that there is no longer any heating of the air by the ground. Discuss the kind of stability that is present in the layer of air immediately above the ground at each time.

7   Suppose the maximum temperature on a summer afternoon is 30°C. What is the approximate temperature at 1 km at that time? If the pressure is 1010 mb at the ground, what is the approximate pressure at 1 km?

8   What is the significance of inversions with respect to air pollution?

# FIVE

## Motions in the Atmosphere. Small-Scale Circulations

### 5.1  Air in motion

We know from common experience that air does not stay still, but moves about, sometimes gently, as a pleasant breeze, sometimes with the gusty violence of a gale or hurricane. It is free to move in all directions, and, indeed, the vertical motions are very important, being responsible largely for the formation of clouds and precipitation. But, ordinarily, horizontal motion predominates. The horizontal motion of air is called *wind*.

The wind direction is the direction *from* which the wind is coming. To remember this sometimes confusing convention the nursery rhyme "The North wind doth blow and we shall have snow" is helpful, at least for those in the Northern Hemisphere. In speed it is of the order of meters per second. We may use the following rough equivalents:

$1$ m s$^{-1}$: light breeze
$5$ m s$^{-1}$: moderate breeze
$10$ m s$^{-1}$: fresh wind
$15$ m s$^{-1}$: gale
$20$ m s$^{-1}$: strong gale
$25$ m s$^{-1}$: storm
$\geqslant 33$ m s$^{-1}$: hurricane
($1$ m s$^{-1} \approx 2$ knots $\approx 2\frac{1}{4}$ mi/hr)

**Figure 5.1**  *Photograph taken from Gemini 12 spacecraft off the west coast of Mexico, November 13, 1966, showing air motions of various types and scales—streaks, waves, and vortices—reflected in the cloud patterns.* [*Courtesy of NASA.*]

Although winds are gusty and variable they form systematic patterns. Some patterns are very large, such as the those formed by trade winds that carried the great sailing vessels of the eighteenth and nineteenth centuries across the Atlantic and Pacific Oceans. An analogous recently discovered phenomenon is the high-level jet stream, which often adds more than 50 m s$^{-1}$ to the speed of jet airplanes in their west-to-east trips across continents and oceans in middle latitudes. Other large systems include the cyclones and anticyclones of middle latitudes with diameters of thousands of kilometers. On a smaller scale are the tropical cyclones (hurricanes and typhoons) ordinarily 100–400 km across.

Still smaller are local storms, including the awesome tornado, which, although extremely strong, is but a few hundred meters in diameter. The swirls of leaves and dust as eddies form when the wind blows past the

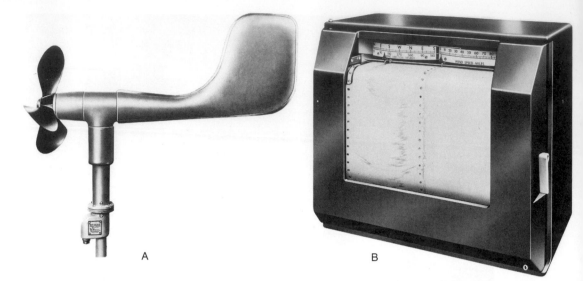

**Figure 5.2** *Wind measuring instruments. A. An instrument in which the anemometer, for measuring wind speed, is combined with a wind vane. The large tail keeps the propellor pointed into the wind. The propellor drives a small electric generator. The output of the generator, which is proportional to the wind speed, is recorded on a strip chart, B, along with a record of the direction toward which the vane is pointing. [Courtesy of Science Associates, Inc.]*

corners of buildings and the convolutions of cigaret smoke reveal very small patterns in the motions of the air. Within the large patterns are imbedded smaller ones, so that the total motion of the atmosphere is an incredibly complex superposition of irregular patterns of all sizes.

To put some semblance of order to this complicated situation we shall examine some of the scales of motion separately, in the hope that the laws that govern them are sufficiently simple and that understanding the parts will help us to comprehend the whole. We start with the factors that lead to air motion.

## 5.2 Newton's laws of motion

The laws of motion discovered by Isaac Newton provide the basis for all analyses of the behavior of moving matter. In a fluid such as air the application of the laws requires some special considerations. Motions in a rotating system such as the earth and its atmosphere likewise introduce differences from those in an *inertial* system. An inertial system is one that is at rest or moving with constant speed in a straight line.

Newton's laws of motion are essentially contained in the expression of his second law, which says that *the rate of change in the amount of motion is proportional to the sum (resultant) of the forces acting and is in the direction the (resultant) force is acting.*

The amount of motion is defined as the product $M\mathbf{v}$ of the mass and the velocity. If a small change $\Delta(M\mathbf{v})$ occurs in a time $\Delta t$, the rate of change is

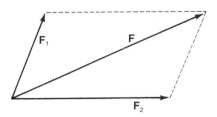

**Figure 5.3** *Addition of vectors.*

the ratio $\Delta(Mv)/\Delta t$. Using this notation we can write Newton's law in the form of an equation:

$$\frac{\Delta(Mv)}{\Delta t} = \mathbf{F}_1 + \mathbf{F}_2 + \ldots = \mathbf{F} \tag{5.1}$$

(Rate of change of motion = Sum of forces = Resultant force)

The expression looks like an ordinary equation but in fact it is a *vector* equation, for each force may have its own direction as well as magnitude. Quantities that have both direction and magnitude (and combine in a certain way) are called *vectors*.[1] When two vectors $\mathbf{F}_1$ and $\mathbf{F}_2$ are added they combine by the *parallelogram* rule, as shown in Figure 5.3.

In ordinary circumstances the mass $M$ does not change. The rate of change of velocity $\Delta v/\Delta t$ is equal to $a$, the acceleration. Equation (5.1) can therefore be written simply as

$$Ma = \mathbf{F} \tag{5.2}$$

that is, the mass times the acceleration is equal to the resultant force.

## 5.3 Forces in a fluid

If we consider a portion of a fluid away from its boundaries, for instance, a parcel of air some distance above the ground, we can see that there are not many forces that can act on it. First, of course, there is the force of gravity, which acts on every mass in the vicinity of the earth. If the parcel carries an electric charge there will be the force of electric attraction (or repulsion), but ordinarily the charges in the air are very small. So, in addition to gravity, the main force acting on a parcel of fluid is the action on it by the fluid around it. This action can be expressed as consisting of two parts, *pressure* and *viscous stress* or internal friction.

If the pressure were the same on all sides of the parcel its effect would cancel out, that is, the resultant force from the pressure forces on all sides would be zero. We know from Chapter 3 that the pressure always varies at least in one direction, the vertical, so that the downward pressure on the top of a parcel of thickness $\Delta h$ is less than the upward pressure on its

---

[1]Vectors are represented in print by boldfaced letters.

bottom. We used the fact that this difference in pressure, resulting in a net upward force, is usually almost exactly balanced by the force of gravity to derive the hydrostatic equation that gives the rate of decrease of pressure with height. Observations, as represented on weather maps, show that the pressure at the same level also varies from place to place, and at any given place as time goes on. This variation gives rise to horizontal net pressure forces.

To summarize the distribution of pressure in the horizontal, isobars (lines along which the pressure is constant, with lower values on one side of the line and higher values on the other) are drawn on weather maps. As we saw in Section 1.3, when this is done relatively simple patterns emerge, with closed isobars enclosing centers of high pressure or low pressure on the surface (sea-level) weather map. At upper levels the isobars tend to be wave-shaped curves circling the earth, with centers of low pressure near the poles. (Instead of isobars at a constant level, contours showing the height of surfaces of constant pressure are used to represent the pressure distribution aloft. The contours of an isobaric surface correspond exactly to isobars for a constant level.)

The rate of variation of pressure represented by the isobars on a weather map is much smaller than the vertical variation. Typically, one would travel a distance of the order of 100 km horizontally to experience a change in pressure of 1 mb, the change one would find in moving vertically about 10 m. The resultant horizontal pressure forces thus may be expected to be much smaller than the vertical, but since they are unopposed by gravity, they may be expected to lead to motions more readily.

It is obvious that the pressure variation results in a horizontal force pushing the air from high to low pressure, tending to produce flow in this direction in the same way that water tends to flow downhill. To see this quantitatively let us consider a parcel of air in the form of a rectangular parallelopiped extending from an isobar drawn for pressure $p_0$ to one drawn for pressure $p_0 + \Delta p$, with the vertical faces parallel and perpendicular to the isobars, as shown in Figure 5.4.

It is clear that the total force acting on face $KLQP$ of the parallelopiped is exactly equal and opposite to that on $MNSR$, since for every point on $KLQP$ there is a point on $MNSR$ where the pressure is the same. Thus the net horizontal force will be the resultant of the forces on the two other vertical faces of the parallelopiped. These forces are directed opposite to each other; their magnitudes are given by the pressure times the area, since pressure is defined as force per unit area. The force on face $KLMN$ is $p_0 A$, where $A$ is its area; the force on $PQRS$ is $(p_0 + \Delta p) A$ in the opposite direction, and the resultant force $F_M$ acting on the mass of air contained in the parallelopiped is

$$F_M = p_0 A - (p_0 + \Delta p) A = - \Delta p \cdot A \qquad (5.3)$$

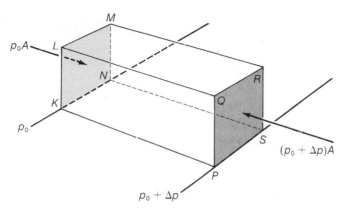

**Figure 5.4** *Illustration of net horizontal pressure force.*

It is desirable to compute the force per unit mass. If the distance $KP = \Delta n$, the volume of the parcel is $A \cdot \Delta n$, and its mass is

$$M = \rho A \cdot \Delta n$$

The magnitude of the force per unit mass [which, by equation (5.2), is equal to the magnitude of the acceleration $a$] is

$$F = \frac{F_M}{M} = -\frac{1}{\rho}\frac{\Delta p}{\Delta n} = -\alpha\frac{\Delta p}{\Delta n} \tag{5.4}$$

The minus sign shows that the resultant horizontal force is opposite to the direction of increasing pressure, that is, in the direction of decreasing pressure. Since isobars are drawn for fixed differences of pressure $\Delta p$ on the weather map, the magnitude of the horizontal pressure force is inversely proportional to $\Delta n$, the distance between isobars. The rate of change of pressure in the horizontal $\Delta p/\Delta n$ is called the *pressure gradient*. The force due to horizontal pressure variations is called the *pressure-gradient force*.

*The pressure-gradient force acts in the direction perpendicular to the isobars towards lower pressure, with magnitude inversely proportional to the distance between isobars.*

In deriving equation (5.3) we should properly have used the average pressure on each of the faces of the parallelopiped, to allow for the variation with height, and in deriving equation (5.4) we should have used the average density. We consider that the vertical dimension of the parallelopiped is taken to be very small, so that the difference between the gradient of the actual pressure and the average pressure can be neglected, and that the density variation is similarly negligible.

To get an idea of how big the forces represented by the isobars on the weather map are let us compute the acceleration corresponding to a typical situation. Suppose 4-mb isobars are 400 km apart on a sea-level chart and the density has its standard value. In cgs units $\Delta p = 4 \cdot 10^3$, $\Delta n = 4 \cdot 10^7$, and $\alpha = 775$. Then

$$a = \frac{775 \cdot 4 \cdot 10^3}{4 \cdot 10^7} = 7.75 \cdot 10^{-2} \text{ cm s}^{-2}$$

If this force were to act for one hour the speed attained would be $3600 \cdot 7.75 \cdot 10^{-2} = 279 \text{ cm s}^{-1} = 2.79 \text{ m s}^{-1}$. Since pressure gradients of this magnitude persist for days, one might expect very large winds to develop unless other forces enter. One such force that is always present is friction, which acts to keep the winds from attaining high velocities much of the time. A more important factor arises from the rotation of the earth and will be discussed in the next chapter.

## 5.4 Causes of horizontal pressure variations

Since the pressure at any point is almost exactly equal to the weight of the air above, per unit area, and since the density depends (inversely) on the temperature, one might expect to have a close correspondence between temperature and pressure, with warm areas having low pressure and cold areas high pressure. In some instances (thermal lows over continental deserts in summer and high-latitude continental highs in winter) this situation dominates, but on the whole conditions are more complicated, with the air column over any given place being cold at some elevations and warm at others. For instance the temperature at 16 km is lower over the equator, where it is in the troposphere, than at 60°N or 60°S, where it is in the stratosphere.

The effect of temperature difference on the pressure-gradient force is most conspicuous in places where differences of terrain produce large differences in the temperature near the ground in short distances over which the conditions at higher levels are practically unchanged. The clearest example is the diurnal variation on clear days at coastal locations. The effect of differential heating and cooling of land and sea produces the sea breeze (from sea to land) during the day and the land breeze at night.

To see how the temperature variations produce the pressure-gradient forces and accelerations in these places consider the schematic drawings in Figure 5.5, in which vertical cross sections through the atmosphere over a flat island and its surrounding ocean are represented. In these cross sections the pressure distribution is represented by isobars. It is assumed that initially the pressure at the earth's surface is the same over the land and the sea. As one goes upward, the pressure drops in accord with the hydrostatic equation, equation (4.4). Where it is cool the pressure drops more rapidly and the distance between isobars is smaller than where it is warm.

Figure 5.5A represents the conditions shortly after sunrise. Over the ocean the temperature at the surface remains low because the heat from the sun penetrates deeply into the water and the air above is relatively cool. Over the island, however, the sun's radiation raises the temperature of the surface, and the air for some distance above it is heated. As a consequence,

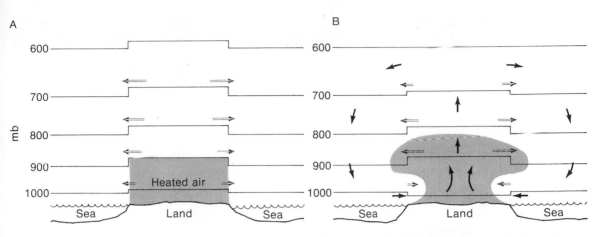

**Figure 5.5** *Schematic development of a sea breeze: double arrows — horizontal pressure gradient forces; single arrows — resultant air motions.*

the isobars, before motion begins, take the form of horizontal lines with vertical steps. Over the sea where the air remains cool, the isobars are close together. In the heated air over the island, they are farther apart. Aloft, the pressure at each level is higher over the island than over the sea.

If the heating were to take place instantaneously there would be no motions at first, only the expansion of the air and the lifting of the isobaric surfaces over the island. This process would give rise to horizontal pressure-gradient forces aloft that tend to push the air outward from the island, as indicated by the double arrows in Figure 5.5A, but would not give rise at first to forces at sea level. After the pressure-gradient forces had acted a while, the accelerations would result in outward movement of the air aloft. The movement from land to sea aloft would cause an accumulation of air at upper levels over the sea that would raise the sea-level pressure there, and correspondingly a depletion of air and decrease in sea-level pressure over the island. These changes in pressure would produce a pressure-gradient force at low levels from the sea toward the island that would accelerate the air from sea toward land, producing the *sea breeze* at low levels. The motions resulting from the seaward forces aloft and the landward forces near the ground are shown schematically by the single arrows in Figure 5.5B.

The actual process is a continuous one, rather than the stepwise process discussed above. The heating, expansion of air, development of pressure-gradient forces, accelerations, and air motions occur together, first slowly as the sun's heating begins to overcome the cooling effects of the night, then more rapidly. The sea breeze reaches its maximum strength in the afternoon, when the heating effect has lasted a long time. Then, as the land cools, it dies down.

At night the reverse effect takes place. The temperature of the land surface falls more than that of the sea surface. This is due to the fact that as

**Figure 5.6** *Visible effect of sea breeze in Florida. Photograph taken from Gemini V spacecraft at 1231 local time, August 22, 1965, looking south along the east coast of Florida. The air over the ocean is virtually cloudless. As it moves onshore it is heated, and cumulus clouds form a short distance inland. [Courtesy of NASA.]*

soon as the sea surface loses some heat it becomes denser than the water below and sinks, so that the cooling is spread through a deep layer of water. The greater cooling of the land surface results in a corresponding decrease in temperature of the air near the ground and a depression of the isobaric surfaces aloft over land relative to the sea. The resulting acceleration of the air from sea to land aloft causes the pressure over land at sea level to rise and produces a seaward pressure gradient there that results in a *land breeze*, that is, a wind from the island towards the sea.

At the coast of a continent the situation corresponds to one-half of the picture over an island. Near the shore the motion is from sea to land during the day and from land to sea at night at low levels, with opposite flow aloft. Far enough inland the temperature is unaffected by proximity to the ocean and there is no influence on the pressure gradient or the winds. Thus at a

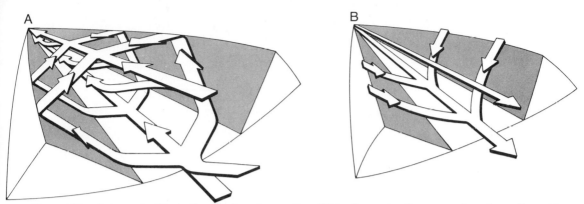

**Figure 5.7** *Schematic illustration of flow in a valley: (A) in the early afternoon when the valley and up-slope winds are occurring; (B) about midnight when the flow is down the valley and the slope. [After F. Defant in* Compendium of Meteorology, *ed. T. F. Malone (Boston: American Meteorological Society, 1951), p. 665.]*

continental coast the sea breeze is felt only 10–20 km inland. Similar effects occur at the shores of large inland lakes.

An effect similar to the sea breeze is produced by differences in land elevation over short distances, producing the valley-mountain wind regime. In this case the air next to the elevated land is heated by the sun more than the air at the same height over the valley. This gives rise to pressure gradients in the way that was described for the sea-breeze situation and produces motion away from the mountain at high levels and toward it near the ground. The flow up the mountains usually is diverted to some extent when there is a U- or V-shaped valley so that the actual flow is a combination of a wind up the valley floor and a *slope wind* directly up the sides of the mountain. At night the winds are reversed, the air cooled by contact with the mountains descending into the valley and flowing downward along its axis. Figure 5.7 shows schematically the flow in a valley (A) in the early afternoon, when the valley and up-slope winds are well developed, and (B) in the middle of the night when the mountain and down-slope winds are near their maximum.

At coastal locations in low latitudes the sea and land breeze pattern is present on almost all days throughout the year, but at higher latitudes the situation is frequently complicated by the superposition of larger-scale wind systems, which may offset the effect of the sea-land temperature contrast. In middle and high latitudes the tendency for a sea breeze may be completely outweighed by the influence of moving large-scale systems. In winter, when the heating by the sun is weak, sea breezes hardly ever occur at these latitudes, and in summer they occur only on days when the prevailing wind is sufficiently light. Even then the sea breeze may start only after enough heating has taken place to overcome a wind from the land produced by a larger-scale system. When the latter situation occurs, the inception of the sea breeze may take the form of a *sea-breeze front,* a sudden invasion of cool air from the sea replacing very hot air. Not infrequently,

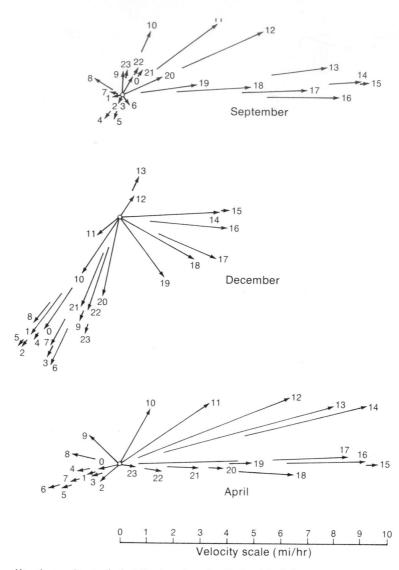

**Figure 5.8**  *Hourly resultant wind at the Los Angeles Federal Building.*

on summer afternoons on the New England coast temperature drops of 20 degrees or more are experienced accompanying the occurrence of sea-breeze fronts.

In locations where mountains are close to the ocean the valley-mountain effect combines with the sea-land influence to produce a more vigorous and more extensive diurnal wind oscillation. Los Angeles, California, is an area in which the combination of the two effects occurs. During the summer the land there cools at night only a little below the sea temperature, and the inland desert and mountains become very hot, so that there is practically no land breeze and a stronger-than-normal sea breeze. In winter, on the other hand, the land becomes colder than the sea for several hours, and the land breeze is stronger and of longer duration than the sea breeze.

Figure 5.8 shows the hourly resultant winds (i.e., winds averaged with respect to direction as well as speed) for three different months at downtown Los Angeles. The general response of the winds to the proximity of the ocean to the west and south, and to the mountains to the north and east, is clearly seen. For instance, in September, which is characteristic of the warm months, the winds are from a westerly direction from 10:00 A.M. to midnight, while from midnight to 8:00 A.M. they are light and variable, but mostly from easterly directions. In December, characteristic of the cold season, the sea breeze begins later and ends earlier, and its strength is somewhat lighter, while the land breeze is correspondingly stronger and of longer duration. The resultant winds for the month shown, December 1947, were apparently affected by the type of large-scale systems already referred to. It appears that a system that resulted in a period of predominantly northerly flow moved through, adding a net north contribution to each hourly resultant and almost eliminating the southerly component of the sea breeze.

Inspection of Figure 5.8 shows that in addition to the general reversal of wind direction between day and night there is a tendency for gradual turning of the wind clockwise (to the right) as the day goes on. This turning is associated with the earth's rotation, which causes moving air to tend to be deflected. This phenomenon will be discussed in the next chapter.

Besides the daily effect of heating and cooling along coasts, in some parts of the world the seasonal variation of the difference in temperature between continents and oceans produces a corresponding seasonal variation in winds. The winds blow from the cooler oceans to the heated continents in summer, and from the centers of cold over the continents toward the oceans in winter. Winds that occur in well-organized seasonal patterns are called *monsoons*, from the Arabic word *mausim*, meaning season. The best-known example is the southwest monsoon of India, which blows from the Arabian Sea across much of the Indian peninsula from June to October, bringing with it general rains that, along the west coast and in the foothills of the Himalayas, are among the heaviest on earth. Similar monsoons are present throughout southern and eastern Asia. For instance, the dominant features of the climate of Vietnam are the southwest monsoon in the summer and the northeast monsoon in the winter. Monsoons are present in greater or lesser degree wherever continents are sufficiently extensive at low latitudes for there to be large areas in which the temperature is markedly different from the adjoining oceans.

There are a number of other local and regional winds. Many of them are due to the local topographic effects on the pressure gradients arising from temperature contrasts produced by large-scale systems. The *mistral*, a northerly wind blowing from the Alps down the Rhone Valley of France toward the Mediterranean Sea, and the *bora*, a similar northeast wind blow-ing from the mountains of Yugoslavia onto the Adriatic, are examples of *fall-winds*, in which the cold air flows downward under the direct action of gravity, like the water in a waterfall. Although the air gets warmer be-cause of adiabatic compression as it descends, it is so cold to begin with that it arrives at the coast as a strong chill wind, which temporarily puts an end to bathing at the resorts of the French Riviera.

The *foehn* of the Alps, the *chinook* of the area east of the Rockies, and the *Santa Ana* of southern California are also due to flow downward from high mountains or plateaus, but in these instances the heating by compres-sion is sufficient to result in abnormally high temperatures and very low humidities. The onset of the chinook is frequently accompanied by rises in temperature of 20°C or more in a few minutes, from below freezing to more than 15°C (59°F), and rapid melting or evaporation of the snow on the ground. Similar rises in temperature accompany the foehn in Europe and the Santa Ana type winds of the west coast of the United States. To the un-pleasant, strong, hot winds and low humidities have been attributed physiological responses called "Foehn sickness." The reality of a well-defined illness due to the foehn is questionable, but there is no doubt that it causes physical discomfort and stress, which may exacerbate already existing ailments.

Occasionally, the foehn, chinook, or Santa Ana winds attain speeds that cause damage to buildings. If fires break out, the low humidities and strong winds cause them to spread rapidly. Thus two of the authors of this book had their homes completely destroyed in a fire that burned thousands of acres of brush in the Santa Monica Mountains of Los Angeles and almost 500 homes before a change in wind and the inflow of moist air permitted it to be controlled.

### Questions, Problems, and Projects for Chapter 5

1   A boat is being propelled across a river at a speed of 4 m s$^{-1}$ perpendicular to the shore. The river is flowing at the rate of 3 m s$^{-1}$. Make a diagram according to the parallelogram rule showing the resultant velocity of the boat.

2   At a point in a fluid the pressure in every direction is the same. In view of this fact, explain how there can be a resultant pressure force that might cause the fluid to accelerate.

3   Make a schematic graph of the variation with time of the pressure at sea level over a tropical island for a 24-hour period and the pressure at a short distance off-shore. Do the same thing for the pressure at a height of about 1 km.

4   If no other forces than the pressure-gradient force were acting, what would be the direction of the winds in relation to the isobars on a weather map? What can you say about the speed of the wind in this case? How would friction with the ground modify this?

# SIX

## Large-Scale Motions

### 6.1 The effect of the earth's rotation

Because horizontal pressure differences arise from the difference in temperature between sea and land at a coast, we would expect the differential heating between equator and poles to give rise to pressure-gradient forces and corresponding circulation of the atmosphere on a large scale. The excess heating at low latitudes would be expected to lead to equatorward motions near the ground and poleward motions aloft. That the winds do not have these directions most of the time is a matter of common knowledge. Near the ground in low latitudes the easterly trade winds are present, while the surface winds at higher latitudes are characterized by the "prevailing westerlies." In the upper troposhere at almost all latitudes the predominant direction of the winds is westerly. The question arises, What is responsible for this zonal (along parallels of latitude) tendency of the winds when the heating should produce meridional flow? The answer lies in the effect of the earth's rotation.

The effect of the rotation of the earth on any moving object is to make it appear as though a force is acting on the object. This apparent force is called the *Coriolis force*, after the French mathematician G. G. de Coriolis (1792–1843) who first treated it quantitatively. The effect had been used qualitatively a century earlier by Hadley to explain the trade winds. The reason for introducing the Coriolis force is that an object moving over the

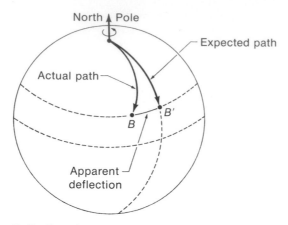

**Figure 6.1** *Deflection of an equatorward projectile.*

earth's surface with constant velocity relative to an absolute frame appears to accelerated relative to a person moving with the rotating earth. Since Newton's law says that if there is an acceleration a force must be acting, we find it convenient to speak of the acceleration as being "caused" by the force that would produce it.

The nature of the apparent acceleration, and the corresponding "force," can be seen in the following way. Suppose that a projectile is fired from the North Pole and aimed at a point *B* at a lower latitude (see Figure 6.1). The earth rotates toward the east, and in the time it takes the projectile to reach *B*, the target that was at *B* will have moved to the position *B'*. An observer on the earth would have expected the projectile to reach the target, since he is not conscious of the earth's rotation. The path he expected would be *PB'*. The actual path of the projectile is *PB*, and thus to the observer the projectile appears to have been deflected to the right. The observer would conclude that there was a force acting at right angles to the motion of the projectile to produce this deflection.

The rotation of the earth toward the east corresponds to counterclockwise rotation when one looks down on the North Pole and to clockwise rotation when one looks down on the South Pole. Consequently, if the projectile were fired from the South Pole instead of the North Pole the deflection would be in the opposite direction. This leads to the following rule: *The Coriolis force acts to the right in the Northern Hemisphere, and to the left in the Southern Hemisphere.*

A similar deflection and apparent force occurs no matter in which direction the projectile moves. If the projectile is aimed parallel to a parallel of latitude (other than the equator) the deflection will also be to the right in the Northern Hemisphere and to the left in the Southern Hemisphere. It is hard to draw a diagram showing this on a sphere. However, the same effect occurs on a rotating disc (e.g., a phonograph turntable) and Figure 6.2 shows how the deviation arises. The projectile is fired from point *A* at the target at point *B*. In the time it takes to get to *B* the target moves to *B'* and

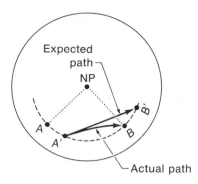

**Figure 6.2**   *Deflection   of   an   eastward   projectile.*

the cannon to $A'$, so the path expected by the observer moving with the disc (earth) is $A'B'$. In following the actual path $AB$ the projectile appears to the observer to be following the curve $A'B$, and thus deviating to the right from the expected path.

Movement in any arbitrary direction can be decomposed into one component along the meridian and one component along the parallel of latitude, for each of which the rotation produces an apparent deflection. Putting the effects of the components back together, it is clear that every such aritrary motion is accompanied by a tendency to be deflected to the right in the Northern Hemisphere and to the left in the Southern Hemisphere. When one is firing an ordinary rifle in target practice or in hunting the effect is too small for one to need to take it into account in aiming. But in aiming long distance projectiles, whether from big cannons or rocket missiles, the Coriolis effect must be included in the ballistic computation.

In large-scale air motion, the effect of the earth's rotation is a dominant factor, and the winds are usually close to the velocity for which the Coriolis force and horizontal pressure-gradient force are in balance. This velocity is called the *geostrophic wind velocity.* In the next sections we shall arrive at quantitative expressions for the Coriolis force and the geostrophic wind velocity.

## 6.2   Quantitative expression for the Coriolis force

To evaluate quantitatively the magnitude of the Coriolis effect we must make use of some additional physical relationships. The first of these is *angular velocity,* which is simply the rate of turning of a rotating body. It is the angle through which the body turns in unit time, and might be expressed in terms of degrees per second, but for convenience in computation it is usually expressed in radians per second, where a radian is $1/(2\pi)$ of a complete circle ($2\pi$ radians $= 360°$).

Since the earth makes one complete rotation in a day (really a sidereal day), its angular velocity $\Omega$ is obtained by dividing $2\pi$ radians by the number of seconds in a sidereal day (86,164.1). This gives $\Omega = 7.29 \cdot 10^{-5}$ radians/sec.

This is the rate at which the earth turns around its axis. At the poles the earth's surface is turning around the vertical at this rate, like a slow-moving phonograph disc. At an arbitrary latitude away from the poles the earth's surface also rotates around the vertical, but more slowly. You can see that it is rotating around the vertical by watching the stars at night, which make partial circuits around the zenith, except at the equator. Thus the rate of rotation around the vertical varies with latitude, from zero at the equator to $\Omega$ at the poles. If $\omega$ is the rate of rotation of the horizontal surface around the vertical at latitude $\phi$, it can be shown that $\omega = \Omega \sin \phi$, where $\sin \phi$ is a trigonometric function that is zero when $\phi = 0$, one when $\phi = \pi/2$ (90°), and 0.7 when $\phi = \pi/4$ (45°).

Next, we must consider the linear velocity of a point a distance $d$ from the axis of a system rotating with angular velocity $\omega$. Since there are $2\pi$ radians in one complete rotation and $\omega$ is the number of radians the system turns in a second, one complete rotation takes $2\pi/\omega$ seconds. The linear distance traveled by a point at a distance $d$ from the axis is $2\pi d$, and thus the linear velocity is $2\pi d \div 2\pi/\omega = \omega d$.

Finally, we need the expression for the distance traveled by a body that is undergoing a constant acceleration, $a$. If the speed is initially zero, at a time $t$ the speed will be $at$. The average speed is $\frac{1}{2}at$, and the distance $s$ traveled is this average speed multiplied by the time, $s = \frac{1}{2}at^2$.

Now we are in a position to derive an expression for the Coriolis force. We return to the treatment in terms of a projectile moving from the axis of rotation. For simplicity we shall consider the motion relative to a rotating disc, which is comparable to the rotation of the horizontal surface at an arbitrary latitude. If (Figure 6.3) it takes a time $t$ for the projectile to move from $P$ to $B$ at a constant speed $v$, we have

$$d = PB = vt$$

To an observer on the rotating disc, who expected the projectile to reach $B'$, the point to which the target originally at $B$ has moved, it appears that the projectile has been deflected a distance $s = BB'$ at right angles to the direction it was aimed. The linear velocity of this deflection is $s/t$, and we have seen above that this velocity is also $\omega d = \omega v t$, so that

$$\frac{s}{t} = \omega v t$$

$$s = \omega v t^2$$

Now, the observer on the rotating disc attributes this deflection to the action of a force (the Coriolis force) that produces an acceleration $a$. For this acceleration

$$s = \frac{1}{2}at^2$$

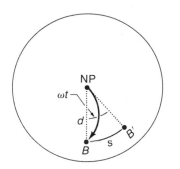

**Figure 6.3**   *Evaluation of the Coriolis force.*

If we set these two expressions for $s$ equal and solve for $a$, we have

$$a = 2\omega v$$

If we substitute for $\omega$ its value for rotation around the vertical at latitude $\phi$, and remember that the acceleration is equal to the force per unit mass, we have the expression for the Coriolis force per unit mass

$$F_c = 2(\Omega \sin \phi) \, v = f \, v \qquad (6.1)$$

where the single letter $f$ is used to represent $2 \, \Omega \sin \phi$, a quantity that occurs repeatedly in meteorological equations. This quantity is referred to as the Coriolis parameter. Table 6.1 gives the value of $f$ at various latitudes, and the Coriolis force corresponding to three speeds at each latitude.

It is interesting to compare the values of the Coriolis force given in Table 6.1 with the example of the magnitude of the horizontal pressure gradient force calculated in Section 5.3. We saw there that if sea-level isobars drawn for every 4 mb are 400 km apart, the pressure-gradient force (per unit mass) is $7.75 \cdot 10^{-2}$ cm s$^{-2}$. It turns out that at 45°, for instance, the Coriolis force would equal the magnitude of this pressure-gradient force if the wind speed is approximately 7.5 m s$^{-1}$. Thus, as suggested at the end of the last section, it is quite possible, for reasonable wind speeds, that the

**Table 6.1**   Variation of Coriolis Parameter and Coriolis Force with Latitude

| Latitude | Coriolis parameter | Coriolis force (cm s$^{-2}$) for speeds of | | |
|---|---|---|---|---|
| $\phi$ (deg) | $f$ (s$^{-1}$) | 5 m s$^{-1}$ | 10 m s$^{-1}$ | 50 m s$^{-1}$ |
| 0 | 0 | 0 | 0 | 0 |
| 15 | $3.8 \cdot 10^{-5}$ | 0.014 | 0.038 | 0.14 |
| 30 | $7.3 \cdot 10^{-5}$ | 0.036 | 0.073 | 0.36 |
| 45 | $10.3 \cdot 10^{-5}$ | 0.052 | 0.103 | 0.52 |
| 60 | $12.6 \cdot 10^{-5}$ | 0.063 | 0.126 | 0.63 |
| 75 | $14.1 \cdot 10^{-5}$ | 0.071 | 0.141 | 0.71 |
| 90 | $14.6 \cdot 10^{-5}$ | 0.073 | 0.146 | 0.73 |

Coriolis force can balance the pressure-gradient force, provided of course that their directions are such that they oppose each other.

Since the Coriolis force always acts at right angles to the direction of motion, it cannot change the speed but only the direction of motion. Acting by itself, it would tend constantly to turn a moving body to the right, and would cause it to go around in circles. The size of the circle would depend on the speed of the moving body (e.g., the projectile), but it turns out that the period of time it takes to go all the way around is independent of the speed, and depends only on the value of $f$, that is, on the latitude. At 30°, for instance, it takes exactly one day; at 45° it takes 0.7 days. This period is called the *inertial period.*

The inertial period is related to the period of the *Foucault pendulum,* which is frequently exhibited in museums to demonstrate the fact that the earth is rotating on its axis. The Foucault pendulum consists of a heavy weight suspended from a mount that permits it complete freedom of motion. The pendulum, when moved from its equilibrium position and let go, will swing parallel to a fixed plane in absolute space. As time passes the earth rotates beneath it, and to the observer moving with the earth, the plane in which the pendulum is swinging appears to rotate slowly. Usually there are markers that are knocked down as the pendulum "rotates," so that the museum visitor can see how far it moves between the time he first sees it and the time he returns after going to see other exhibits. The period of the Foucault pendulum, called a *pendulum day,* is exactly twice the inertial period.

We can now see why the sea breeze, as illustrated in Figure 5.5, turns with time. When it starts (about 9:00 A.M.) air reaching the observing station has been moving in response to the sea-land pressure-gradient force for less than an hour, and thus the Coriolis force will have deflected its direction very little. But by 3:00 P.M. the air reaching the station may have been moving approximately 6 hours, and the Coriolis force would turn it almost one-quarter of the way around (90°), until it parallels the coast. While the presence of other forces, including friction, complicates the situation, it will be seen in the figure that the actual turning of the sea breeze at Los Angeles (latitude 34°) is not far from the rate indicated by consideration of the Coriolis force.

## 6.3 The geostrophic wind

Let us now consider a situation in which initially there are horizontal variations of pressure, as represented by the isobars in Figure 6.4, but no motion. A parcel at point 1 would be acted on by the pressure-gradient force, represented by the double arrow marked $\mathbf{F}_p$, which would accelerate it, displacing it to point 2. At point 2 it would still be acted on by $\mathbf{F}_p$ (which we assume to be constant, for simplicity), but because it is moving it would be acted on also by a Coriolis force $\mathbf{F}_{c2}$, tending to deflect it to the right to

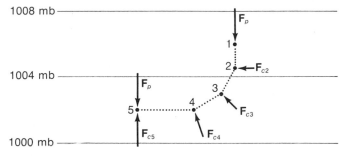

**Figure 6.4**  *Adjustment of wind to geostrophic wind in uniform pressure gradient.*

point 3. By the time it reaches point 3 it is moving faster, and thus $F_{c3}$ is larger and bends the path still more. When it reaches point 5 it is moving parallel to the isobars, so that $F_{c5}$ points toward high pressure, exactly opposite to $F_p$, which is always perpendicular to the isobars and points toward low pressure. If at this time the speed would have the right value so that $F_c$ has the same magnitude as $F_p$ (i.e., $F_c = -F_p$), the parcel would no longer be accelerated, but would move with constant speed along the isobars.

When the air is moving in this fashion, so that the Coriolis force is exactly equal and opposite to the pressure gradient force, it is said to be in *geostrophic balance,* and the wind velocity for which this is true is called the *geostrophic wind.*

It can be shown mathematically that if the initial situation were as postulated above, with no other forces acting, and if the air motion did not produce changes in the pressure distribution, at the time that the motion of the air parcel became parallel to the isobars it would be moving so fast that $F_0$ would be considerably greater than $F_p$ and thus would be accelerated toward higher pressure. It would then be slowed down by the pressure force acting opposite to the direction of motion. The result would be that instead of reaching the equilibrium shown by point 5 in Figure 6.4, it would oscillate around it. It would "try" to attain the equilibrium but would always overshoot it.

In actuality the air does not start from rest, but is always in a state much closer to geostrophic balance. In effect, the wind is continually attempting to adjust itself to a balance between the pressure-gradient force and the Coriolis force. Thus, with the exception of locations near the ground, where frictional forces are important, and places where the isobars are strongly curved (which will be discussed in the next section), the geostrophic wind is a good approximation to the real wind. The importance of this lies in the fact that as a consequence the wind flow can be deduced from the isobaric direction and spacing.

The direction of the geostrophic wind is parallel to the isobars, with low pressure to the left in the Northern Hemisphere (to the right in the Southern Hemisphere). The relation between the isobaric spacing and the geostrophic wind speed is obtained by setting the magnitude of the Coriolis force, given in equation (6.1), equal to that of the pressure-gradient force, given by equation (5.4). If $v_g$ represents the geostrophic speed,

$$v_g = \left| \frac{\alpha}{f} \frac{\Delta p}{\Delta n} \right| \tag{6.2}$$

The geostrophic wind speed varies —

1 with pressure gradient, or inversely with the distance between isobars;
2 with specific volume, or inversely as the density;
3 inversely with the Coriolis parameter, or inversely with the sine of the latitude.

For the isobaric spacing of 100 km/mb and specific volume of 775 cm³/g for which we computed $F_p$ in Section 5.3, the geostrophic wind speed would be the following:

10.6 m s⁻¹ at 30° latitude

7.5 m s⁻¹ at 45° latitude

6.2 m s⁻¹ at 60° latitude

At low elevations friction with the ground causes considerable deviations from the geostrophic wind. We shall discuss the effect of friction later. At upper levels friction is negligible and the wind blows almost exactly parallel to the isobars with its speed given approximately by equation (6.2).

Drawing isobars on a weather map thus is a good way to summarize the wind distribution some distance above the ground, since isobars are approximate *streamlines,* showing the direction in which the air is moving, and their spacing indicates the wind strength, light where they are far apart and strong where they are close together. As we shall see, the direction and speed of the winds at low levels also can be inferred from the isobars.

## 6.4 Curved flow. The gradient wind

As we have seen in the examples of the weather map in Section 1.3, the isobars are not straight lines but curves that sometimes enclose centers of high or low pressure. In order to remain parallel to the isobars the wind would have to follow curved paths and thus be accelerated. For curvature of the flow around low pressure, the pressure-gradient force must be greater than the Coriolis force, and the opposite is true for curvature around high pressure. Since the Coriolis force depends on the wind speed, we see that the speed of wind corresponding to flow following curved isobars is different than for straight isobars. For the same spacing of isobars at the same latitude the wind speed is less for flow curved around low pressure than for straight flow; for flow curved around high pressure it is greater than for straight flow.

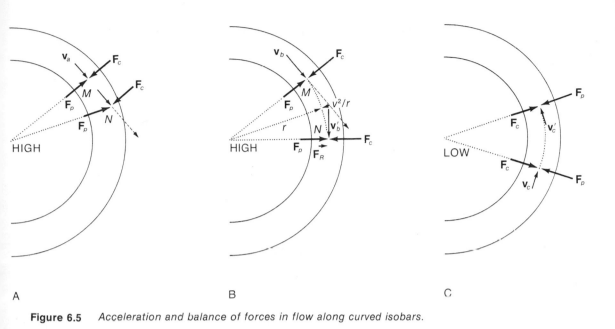

**Figure 6.5** *Acceleration and balance of forces in flow along curved isobars.*

To see why this is true, consider a situation with circular isobars curving around a high-pressure center, as illustrated in Figure 6.5A and B.

In diagram A the air at point $M$ is assumed to be moving with the geostrophic velocity $\mathbf{v}_a$ corresponding to the direction and spacing of the isobars. After a time it will have moved in a straight line to the point $N$. At that point the direction of the isobars is different, so $\mathbf{F}_p$, which points away from the high center, will no longer exactly balance $\mathbf{F}_c$. In the diagram it is seen that the unbalanced resultant force would lead to an increase in speed of the air parcel, thus increasing $\mathbf{F}_c$. In turn, the increased Coriolis force would lead to an acceleration to the right, toward the high center.

In diagram B a larger velocity $\mathbf{v}_b$ is assumed, so that $\mathbf{F}_c$ is sufficiently larger than $\mathbf{F}_p$ to push the air constantly toward the high-pressure center just the right amount to keep the air moving along the circular isobar.

The Coriolis force, $\mathbf{F}_c$, which is always perpendicular to $\mathbf{v}$, will remain exactly opposite to $\mathbf{F}_p$, and since both $\mathbf{F}_c$ and $\mathbf{F}_p$ are therefore at right angles to the direction of motion at each instant, there is no change in speed. Only the direction changes, but a change in direction of motion is an acceleration, as well as a change in speed. Acceleration toward a center (in this case the center of the HIGH) is called *centripetal* acceleration. Whenever a body moves in a curved path it is undergoing centripetal acceleration. It can be shown that if the radius of curvature of the path is $r$, the amount of the acceleration is $v^2/r$.

The above discussion shows that when the speed is larger than that required for geostrophic balance, so that $\mathbf{F}_c$ is larger than $\mathbf{F}_p$, there will be a centripetal acceleration that produces curved flow around high pressure.

Sometimes, rather than speak of the centripetal acceleration associated with curved motion, reference is made to the *centrifugal force*. This is the apparent force outward from the center that the curved motion appears to exert to balance the forces producing the centripetal acceleration.

The situation in which the isobars are curved around a low center is shown in diagram C. In this case $F_p$ is directed toward the center, and in order that the air may move exactly along an isobar around the center, $F_c$ must be smaller than $F_p$. The speed for flow at constant speed around a low center is thus *less* than the geostrophic speed.

The illustrations in Figure 6.5 are for the Northern Hemisphere, for which the Coriolis force acts to the right of the direction of motion. Inspection of this figure leads to the rule: *In the Northern Hemisphere air normally flows clockwise around high-pressure centers and counterclockwise around low-pressure centers.* This statement is true since normally the winds are as close to geostrophic balance as they can achieve. In the Southern Hemisphere the normal flow is exactly the opposite: clockwise around low-pressure centers and counterclockwise around high centers.

The normal flow around low-pressure areas is called *cyclonic flow*, and the systems consisting of low pressure centers and the accompanying cyclonic wind systems are called *cyclones*. *Anticyclones* correspondingly refer to the high centers and accompanying anticyclonic flow.

It is important to remember:

1  *Cyclonic flow is counterclockwise in the Northern Hemisphere and clockwise in the Southern Hemisphere.*
2  *Anticyclonic flow is clockwise in the Northern Hemisphere and counterclockwise in the Southern Hemisphere.*

The wind illustrated in diagrams B and C of Figure 6.5 is called the *gradient wind*. The gradient wind is the wind that would flow along curved isobars with no change in speed. From the preceding discussion we see that the speed of the gradient wind is greater than the geostrophic wind speed for anticyclonic flow, and less for cyclonic flow. The quantitative relationship can be calculated from the balance of forces shown in the diagrams.

If we define the radius of curvature $r$ as positive for cyclonic flow and negative for anticyclonic flow, we can write a single equation for the two cases:

$$F_p - F_c = v_G^2/r \tag{6.3}$$

where $v_G$ is the gradient wind speed. If we note from equation (6.2) that $F_p = fv_g$, where $v_g$ is the geostrophic speed for straight isobars having the same spacing, and from equation (6.1) that if the velocity is $v_G$, $F_c = fv_G$, we get

$$fv_g - fv_G = v_G^2/r \tag{6.4}$$

For clarity in interpreting the result it is convenient to divide both sides of the equation by $v_G{}^2$, solve the resulting quadratic equation in $1/v_G$, and then take the reciprocal. The result is

$$v_G = \frac{2\,v_g}{1 + \sqrt{1 + 4\,v_g/fr}} \tag{6.5}$$

The plus sign is chosen for the radical to select the solution that reduces to the geostrophic wind speed when $r$ increases indefinitely, corresponding to straight flow.

This equation, too, shows that for flow around low pressure ($r$ positive) $v_G < v_g$, while for flow around high pressure, $v_G > v_g$. It also shows that for anticyclonic flow ($r$ negative) there is a limit on the pressure gradient close to the center if the flow is to follow the isobars, since if $v_g > -fr/4$, $v_G$ would not be real. For anticyclonic gradient flow, the geostrophic speed, and thus the pressure gradient, must approach zero as the center is approached. There is no such limitation for the pressure gradient near low-pressure centers.

To get some idea of the amount of difference between the gradient wind and the geostrophic wind, let us consider again the example of 4-mb isobars 400 km apart, for which we found $v_g = 7.5$ m s$^{-1}$ at 45° latitude, where $f = 10.3 \cdot 10^{-5}$. For $r = 750$ km (cyclonic flow in a circle with a radius of 750 km) we have (using units of meters and m s$^{-1}$)

$$v_G = \frac{2 \cdot 7.5}{1 + \sqrt{1 + 4 \cdot 7.5/(10.3 \cdot 10^{-5} \cdot 7.5 \cdot 10^5)}} = 6.9 \text{ m s}^{-1}$$

If the flow were anticyclonic ($r = -750$ km)

$$v_G = \frac{2 \cdot 7.5}{1 + \sqrt{1 - 4 \cdot 7.5/(10.3 \cdot 10^{-5} \cdot 7.5 \cdot 10^5)}} = 8.4 \text{ m s}^{-1}$$

Sometimes the gradient wind is defined in terms of the curvature of the actual path of the air rather than the isobars, since the path may deviate from the isobars because of changes of the pressure field with time, or for other reasons.

In addition to the deviation of the air trajectories from the isobars, other accelerations are frequently as large as the centripetal acceleration, except in cases of large curvature. Except near the centers of hurricanes and other intense cyclonic storms, these other deviations may render the gradient wind speed no better an approximation of the actual speed than the geostrophic speed.

## 6.5   The effect of friction

In deriving the expressions for the speed of the geostrophic wind and the gradient wind we have ignored friction. Sufficiently far from the earth's

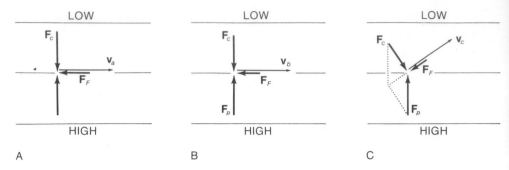

**Figure 6.6** *Effect of friction on surface winds.*

surface it is small enough to be ignored, but close to the ground the frictional force cannot be neglected. A rough value for the thickness of the layer of frictional influence is 1 km. Above this height deviations from the geostrophic wind or the gradient wind are usually small. The deviation due to friction at the ground is large and decreases upward.

Near the ground (say, at the anemometer level, which is usually 10 or 15 m above the ground), the wind is reduced and inclined towards low pressure by an amount that depends on the nature of the surface. Over the sea the surface wind has a speed that is about two-thirds of the gradient speed and a direction that makes an angle of about 15° to 25° to the isobars. Over fairly rough ground the surface wind may be one-half of the gradient wind speed or less and inclined about 30° to 40° from the isobars.

A crude interpretation of the effect of friction, which qualitatively explains why the wind blows toward lower pressure with reduced speed when friction is acting, is shown in Figure 6.6. The force due to friction with the ground is assumed to act opposite to the surface wind and be proportional to it. If, as in diagram A, we start with a situation in which the surface wind is blowing along straight isobars with geostrophic speed, the pressure-gradient force $\mathbf{F}_p$ and the Coriolis force $\mathbf{F}_c$ will just balance each other, but friction with the ground will cause a force $\mathbf{F}_F$ to act on each parcel of air in the opposite direction to the direction of the wind. This unbalanced force will decelerate the parcel, creating the situation represented in diagram B, where $\mathbf{F}_c$ is smaller, corresponding to the reduced wind speed. The excess of $\mathbf{F}_p$ over $\mathbf{F}_c$ pushes the parcel toward lower pressure. Balance is achieved when, as shown in diagram C, the direction and speed of the wind are adjusted so that the resultant of $\mathbf{F}_c$ and $\mathbf{F}_F$ is exactly equal and opposite to $\mathbf{F}_p$.

The effect of friction with the ground is transferred through the air by the irregular motions of particles of fluid. As mentioned in Section 5.3, the force exerted on a parcel of fluid by the fluid around it consists of two parts, the pressure and the *viscous stress*. The pressure is the force pushing perpendicular to the boundaries of the parcel. The stress is the drag that the fluid around it exerts tangential to the boundaries of the parcel. This

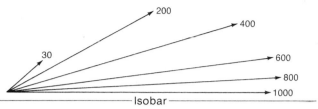

**Figure 6.7** *Variation of wind with height due to surface friction. Arrows represent wind at heights (in meters) given by numbers at their points.*

viscous stress, or internal friction, exerts a force on the boundaries of the parcel proportional to the variation of velocity with distance across the boundaries. The variation of velocity with distance is called the *shear* of the wind, and the resultant horizontal force on a parcel bounded by plane surfaces above and below is the force due to the shear across the upper surface minus the force due to the shear across the lower surface.

If the molecular motions were the only irregular motions causing viscous stress, the effect of friction with the ground would be felt through a very thin layer. The roughness of the ground and an innate instability of the flow cause larger irregularities, or turbulence, which is responsible for the transfer of the effect of the ground up to considerable heights. When heating at the ground causes convective motions the thickness of frictional influence is further augmented, and if convective clouds, showers, and thunderstorms develop the entire troposphere may be affected.

The variation of wind with height in the layer of frictional influence is of the type shown in Figure 6.7, in which the arrows represent the wind velocity at the heights (in meters) indicated by the numbers at their points. The speed of the wind increases with height, and its direction turns to the right (in the Northern Hemisphere) until, at the top of the layer, it approximates the geostrophic velocity.

This change of wind with height is always present in the layer next to the ground. At greater heights there is frequently further change, due to the fact that the pressure systems vary with height. This variation will be examined in the next section.

## 6.6 Variation of pressure gradients with height. The thermal wind

Application of the hydrostatic equation [equation (4.4)] shows that where the temperature is high the pressure decreases slowly with height, and where it is low the pressure decreases rapidly. At higher levels the pressure gradient is the result of the pressure gradient at sea level plus a contribution due to the horizontal gradient of the average temperature in the layer up to the particular level.

The change in pressure gradient as one goes upward that is due to the horizontal variation of temperature is demonstrated in Figure 6.8. This

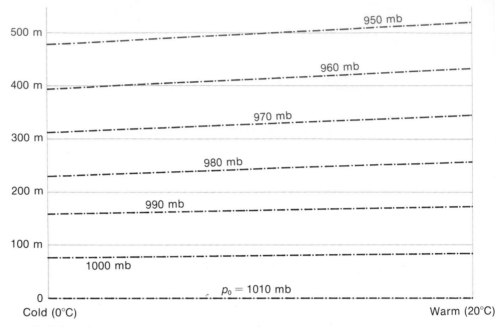

**Figure 6.8** *Variation of slope of pressure surfaces with height.*

figure is a vertical cross section in which the temperature increases from left to right at all levels. For simplicity it is assumed that the pressure has a constant value of 1010 mb at sea level. On the left, where the air is cold, the pressure decreases more rapidly than on the right, where the air is warmer. Thus the distance to the position where the pressure is 1000 mb is smaller on the left than on the right, and, similarly, for the distance from 1000 to 990 mb, and so forth. The slope of the isobars in the cross section increases with height, and so does the pressure difference at constant elevations. Thus the horizontal temperature gradient leads to a pressure gradient at upper levels.

Corresponding to the change in pressure gradient with height, there will be a change in the geostrophic wind. This change in geostrophic wind with height because of temperature variation is called the *thermal wind*. The magnitude of the thermal wind is proportional to the gradient in average temperature. In fact, the formula for it is very similar to the geostrophic wind formula [equation (6.2)]. It may be written approximately

$$v_T = \left| \frac{g}{fT} \frac{\Delta T}{\Delta n_T} \right| \Delta H$$

where $v_T$ is the magnitude of the change in the geostrophic wind vector in going upward a distance $\Delta H$, and $\Delta n_T$ is the distance between isotherms (of average temperature through the layer $\Delta H$) drawn for intervals of $\Delta T$. The other terms have been previously defined. The direction of the thermal wind is related to the isotherms in the same way as the geostrophic wind

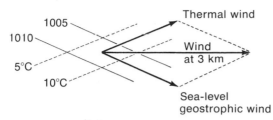

**Figure 6.9** *Illustration of thermal wind.*

is related to the isobars: *The thermal wind "blows" along the isotherms with low temperature to the left in the Northern Hemisphere (to the right in the Southern Hemisphere).* As an indication of the magnitude of the thermal wind, at 45° latitude a horizontal temperature gradient of 1°C/100 km corresponds to a thermal wind of about 10 m s$^{-1}$ between sea level and a height of 3 km.

Since the isotherms in general are not parallel to the isobars the direction of the thermal wind will usually be different from that of the geostrophic wind. To estimate the wind at upper levels the thermal wind must be added vectorially to the geostrophic wind at the lower level. Figure 6.9 shows an example of the thermal wind and the resulting wind at an upper level.

In addition to being proportional to the temperature gradient (and therefore inversely proportional to $\Delta n_T$, the distance between isotherms), the thermal wind is proportional to the thickness $\Delta H$ of the layer through which the change is being computed. If the isotherms do not change much in direction with height, as is often true through most of the troposphere, for large enough $\Delta H$, that is, at high levels, the thermal wind forms the major part of the wind. This means that as you go up the wind usually becomes more and more nearly parallel to the isotherms, up to the tropopause. Correspondingly, the isobars on upper-level maps (and the isobaric contours on constant-pressure maps) become approximately parallel to the isotherms.

The temperature variation in the horizontal also affects the position and nature of high- and low-pressure centers at higher levels. For instance, if a high center is colder than its surroundings the pressure will decrease more rapidly at the center than around it, and at a sufficient height the pressure at the position of the high at sea level will no longer be higher than in the surrounding area. In other words *cold highs weaken and disappear with height, while warm highs surrounded by colder air become more intense.* Similarly, *warm lows are shallow and cold lows extend to great heights.* If a center has a temperature gradient across it, it will tilt with height: *low-pressure centers tilt toward low temperatures; high-pressure centers toward high temperatures.* The configuration of the pressure field at upper levels may thus be estimated from the temperature field at sea level. This procedure will be illustrated in Section 11.5, in which maps of the pressure distribution at sea level and at 500 mb are compared.

## Questions, Problems, and Projects for Chapter 6

1  The moon rotates once every lunar month. Is the Coriolis force at a given latitude on the moon larger or smaller than at the same latitude on earth? By how much?

2  If the wind is 10 m s$^{-1}$ at the equator, what is the magnitude of the Coriolis force there? Can the wind be geostrophic? Why? What does this imply with respect to the relation between wind direction and isobars?

3  Suppose that you find that at a particular place, say, San Francisco, the isobars at the 3-km level are the same distance apart and in the same direction as the isobars on the sea-level weather map. Would you expect the geostrophic wind speed to be the same at 3 km as at sea level? Greater? Less? Why?

4  Why does the wind near the ground blow toward low pressure while the wind aloft blows almost exactly parallel to the isobars? Does the same rule apply in the Southern Hemisphere?

5  Since the Coriolis force tends to deflect moving air parcels to the right in the Northern Hemisphere, why do the winds around a low-pressure area turn to the left (counterclockwise)?

6  a  Make two drawings showing the surface wind direction around (i) a low-pressure center and (ii) a high-pressure center in the Southern Hemisphere.

b  Is the gradient wind speed around a low-pressure center in the Southern Hemisphere greater or less than the geostrophic wind speed? Draw a diagram that shows why your answer is right.

7  A hurricane is an intense rotating storm originating at low latitudes, where the Coriolis force is small. If the Coriolis force can be neglected near the center, write the equation for the relation between wind speed and the pressure gradient, neglecting friction. What would the wind speed be for concentric 5-mb isobars 10 km apart at a distance of 30 km from the center if $\alpha = 0.800$ m$^3$/kg?

# SEVEN

# The General Circulation of the Atmosphere

## 7.1    Deflection of meridional circulations

At the beginning of Chapter 6 the difference between the observed large-scale wind systems and the flow patterns to be expected because of greater heating at low latitudes than near the poles was pointed out. Instead of the meridional circulation that would be expected, with heated air rising near the equator, flowing poleward aloft, descending while being cooled in polar regions, and then flowing equatorward near the ground, the observed flow is predominantly zonal, easterly in low latitudes and westerly at high latitudes near the ground, and westerly at almost all latitudes aloft.

We can now see the reason, in terms of the action of the Coriolis force, why the winds are predominantly zonal on the average. The air moving initially along the meridians in response to the pressure gradient arising from the heating—poleward pressure-gradient force aloft and equatorward pressure-gradient force near the ground—would be deflected to the right in the Northern Hemisphere and to the left in the Southern Hemisphere until a balance was reached. If the pressure force were the simple one expected from the distribution of heating and cooling, the balance would require that the winds be easterly near the ground at all latitudes, and westerly aloft. Since the observed surface winds are both easterly (at low

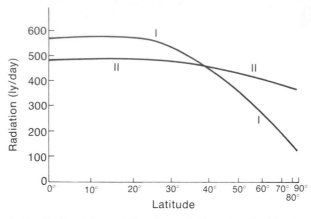

**Figure 7.1**  *Latitudinal variation of (I) solar radiation absorbed by the earth and atmosphere and (II) long-wave radiation leaving the atmosphere. [From H. G. Houghton, "On the Annual Heat Balance in the Northern Hemisphere," J. Meteorol. 11 (1) (1954):7.]*

latitudes) and westerly (at high latitudes), it is clear that the simple discussion used so far is only part of the explanation.

A further deficiency of the explanation lies in the fact that if the winds were completely zonal there would be no transfer of heat from low latitudes to high latitudes. The excess heating at low latitudes would result in a continuous increase in temperature there. A corresponding decrease in temperature would occur at high latitudes. In the next section we shall analyze in more detail the need for meridional (south-north) transfer of energy.

## 7.2  Radiation balance

We would expect that the earth, taken as a whole, is approximately in radiative equilibrium; that is, that about the same amount of energy is being radiated by it to space as it is receiving from the sun. The reason we expect this is that if more were being received than radiated away, the average temperature of the earth would rise. The amount radiated away increases with temperature, and thus the outgoing radiation would increase until the temperature had risen to the point at which the outgoing and incoming radiation were equal. The earth and its present atmosphere have been in existence long enough for this balance to have been attained. (In recent years a slight departure from this balance may have been produced by the increase in the amount of carbon dioxide and particulate matter in the atmosphere.)

While the radiation is approximately in balance for the earth as a whole, for any particular latitude the incoming and outgoing radiation are not equal, even when averaged for the entire year. Figure 7.1 shows these quantities as computed approximately, using estimates of the average cloudiness, water-vapor content, and temperature. Curve I represents the year-round average radiation absorbed by the earth and atmosphere, and Curve II the radiation going out from the earth and atmosphere to space, for various latitudes. At low latitudes Curve I is higher, showing excess energy received from the sun, and at high latitudes Curve II is on top, representing a net loss of energy.

The total radiation for the earth (or, rather, one hemisphere) is proportional to the area under each curve. The areas under Curve I and Curve II are equal, showing that the yearly income and outgo are in balance.

The departures from balance at particular latitudes are considerable. For instance, at 20° the short-wave radiation received from the sun and absorbed is about 24 MJ/m² (570 langleys) per day, but the outgoing long-wave radiation is less than 21 MJ/m² (500 langleys) per day. At all latitudes equatorward of 38°, where the curves cross, energy tends to accumulate on the average, while poleward a deficit occurs. It is this imbalance that would produce meridional circulations to carry the excess heat to the regions of deficiency if the Coriolis force permitted, except for the small amount that can be carried by ocean currents. George Hadley, in 1735, used the interaction of the energy imbalance and the deflecting influence of the earth's rotation to explain the easterly tradewinds. His explanation is still regarded as valid for the circulation at low latitudes. In the next section a discussion of the Hadley circulation will be presented and the reason why it cannot apply to the air flow at higher latitudes will be explained.

## 7.3 Hadley cells and angular momentum

Figure 7.2 shows schematically the circulations that should arise from excess heat at low latitudes and deficiencies at high latitudes. The equatorward flow near the earth's surface would produce easterly winds; the poleward flow aloft would produce westerly winds. Because of friction the geostrophic balance would not be attained near the ground, and the winds would be northeasterly in the Northern Hemisphere and southeasterly in the Southern Hemisphere. There would be a net flow toward the equator at low levels, and continuity of mass would require a poleward current aloft superposed on the westerlies. The meridional circulations that result (with speeds much less than those of the easterly and westerly components) are called the *Hadley cells*.

There are two reasons why the general circulation cannot consist simply of Hadley cells (one in each hemisphere). One has to do with the conservation of angular momentum of rings of air as they move southward and northward, and the other with the conservation of angular momentum of the earth-atmosphere system.

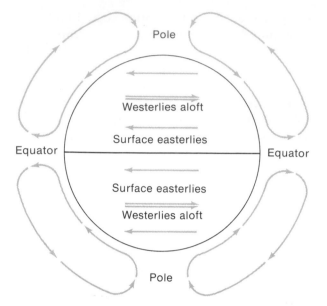

**Figure 7.2** *Hadley model of general circulation: single arrows—winds near ground; double arrows—winds aloft.*

Applied to rotating bodies, Newton's laws of motion say that the rate of change of angular momentum (amount of rotational motion) of a body is equal to the sum of the torques (twisting forces) acting. A torque is expressed as a force acting at right angles to a radius from a center of rotation. An example of a torque is the twisting action used to tighten a bolt with a wrench. The force is applied perpendicular to the wrench handle. In the absence of torques the angular momentum is constant. Since no external torque is exerted on the earth-atmosphere system its angular momentum is conserved.

Angular momentum is measured by the product $Mvr$, where $r$ is the distance of the mass $M$ from the center of rotation and $v$ is the linear velocity due to the rotational motion. If no torques are acting on an isolated mass we have the result

$$Mvr = \text{constant}$$

so that if $r$ decreases, $v$ increases, since $M$ is constant. If a body consists of masses at different distances we must add up the angular momentum of all its component masses.

A familiar example of conservation of angular momentum is that of a figure skater twirling first with her arms extended, then with her arms pulled in toward her body. As her arms move in, more of the mass becomes concentrated close to her center of rotation and she consequently rotates faster.

If mass remains the same and distance decreases, then to conserve angular momentum, the speed must increase.

Another example is that of an object being swung around on the end of a string. Let it rotate at a constant rate and then pull the string inward so that the distance from the axis of rotation decreases. Since the angular momentum remains the same, the speed of the object increases.

We shall now apply the principle of conservation of angular momentum to the rings of air in the Hadley circulation shown in Figure 7.2. If we assume that the air rising near the equator is stationary relative to the earth, in absolute motion it has a speed $v$ of about 465 m s$^{-1}$ or 1,000 mi/hr from west to east. As it moves poleward at upper levels its distance $r$ from the earth's axis decreases, and in order that $vr$ may be constant $v$ must increase correspondingly. When it reaches 45°, the distance is reduced to 0.707 of its original value. To maintain the constancy of the product $vr$ the absolute velocity $v$ must increase to 658 m s$^{-1}$. The speed of the earth's surface at 45° latitude, on the other hand, is smaller by the same proportion as $r$, 329 m s$^{-1}$. The air would thus be moving relative to the earth with a speed of 329 m s$^{-1}$ (737 mi/hr). While such extremely high wind speeds (and the corresponding pressure gradients) could conceivably exist, the theoretical analysis shows that they would result in great instability, in the sense that any small disturbance from the zonal flow would grow and cause a breakdown in the zonal circulation. Only in the low latitudes, where the distance reduction is small enough, does conservation of angular momentum result in speeds of zonal flow that are relatively stable so that the zonal motion so produced can persist. For this reason the Hadley circulation is limited to low latitudes.

The other reason why the simple pressure pattern that the Hadley circulation requires cannot exist over the entire earth is concerned with the effect of friction between the atmosphere and the earth's surface. The pattern at the surface, with low pressure at the equator and high pressure at the poles, would require easterly winds at all latitudes for approximate geostrophic balance. Friction between the easterly winds and the earth's surface would produce a torque, with the atmosphere tending to slow down the earth and the earth tending to speed up the atmosphere. This condition could not persist. Since there is nothing for the air to "hold onto" while pushing on the earth to slow it down, it would quickly reach equilibrium in which it would move with the earth, with no wind at all. Thus surface winds in one direction over the entire earth cannot exist; westerlies at some latitudes are necessary, in addition to easterlies at others, to maintain a constant average flow pattern. Since friction with the earth would tend to slow down both easterlies and westerlies, there must be a process by which westerly momentum is transferred from the easterlies of low latitudes to the westerlies of temperature latitudes in order to keep both of these currents going.

Since the Hadley cells are confined to low latitudes, the transfer of the excess heat across middle latitudes towards the poles likewise requires another process. The process that is operative involves the action of smaller-scale circulation elements, vortices and waves, which will be discussed in the next section.

## 7.4 Transport of heat and momentum by waves and vortices

We have seen that a simple cellular flow pattern consisting of one Hadley cell in each hemisphere cannot carry out the function of transporting heat all the way from the equator to the poles without giving rising to unstable conditions that would cause a breakdown in the zonal circulation and a poleward limit to the Hadley cells. This breakdown takes the form of waves in the zonal flow and the development of complete swirls—cyclonic and anticyclonic vortices. These disturbances have such character that they transfer heat and momentum north and south even though the net meridional movement of air across parallels of latitude averages out to be approximately zero at each level. That is, if at some place and time there is a northward flow, at another place and time there is a southward flow of the same amount of air at the same level.

To see how there can be a transport of heat and momentum by north-south movements even though the flow of air cancels out, suppose that the northward-moving air in the Northern Hemisphere is warmer than the air moving southward at the same latitude at every level. Less heat would be carried southward across the parallel of latitude than would be carried northward, and there would be a net transport of heat from south to north. If, on the average, the absolute zonal speed of the northward-moving air was greater (less strong easterly or stronger westerly components of wind) than that of the southward-moving air, there would similarly be a transfer northward of westerly momentum. Deviations of this type from the average for the latitude are to be expected. In the Southern Hemisphere, of course, it is the southward flow that is on the average warmer and more westerly than the northward. Thus the transfers necessary to carry poleward the excess heat received at low latitudes and to maintain the easterlies and westerlies near the ground (and the westerlies aloft in higher latitudes) can be achieved.

The details of how the transfer of heat and momentum occurs and why the disturbances from the mean flow have the character that is observed have been studied in model experiments in the laboratory and by the use of high-speed electronic digital computers. These studies have helped us understand the large-scale behavior of the atmosphere.

## 7.5 Experiments on the general circulation

While the full complexity of the atmosphere cannot be simulated in the laboratory, certain essential features of the general circulation can be reproduced by simple experiments called "dishpan" experiments, in which a pan of water is heated on a rotating table. The general configuration of the experiment is shown in Figure 7.3. The pan is heated at the rim (equator)

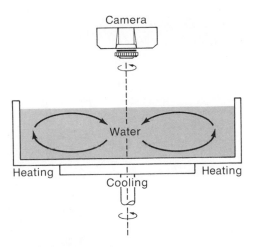

**Figure 7.3** *Diagram illustrating apparatus used in "rotating dishpan" experiment.*

and cooled at the center (pole). Particles of dust suspended in the water enable its motions to be seen, and a motion-picture camera rotating at the same rate as the pan records how the motions would look to an observer on the pan (earth). If the pan is not rotating, the motion of the water in the pan is just the meridional motion of the Hadley circulation shown in the figure. When the pan is rotating zonal motions develop, and the character of the flow changes with the rate of rotation.

For very slow rotation the flow consists of a simple zonal motion on which a slow Hadley circulation is superimposed. As the rate of rotation increases, the simple flow breaks down into a series of waves and vortices. The size and number of the eddies thus produced depend on the relation of the heating rate to the rate of rotation. For small heating rates, which produce small temperature differences, and for large rotation rates the eddies are numerous; for large temperature differences and small rotation rates fewer large eddies or waves occur. Figure 7.4 is a photograph of a rotating dishpan experiment in which the flow has broken down into a large number of waves and vortices. It is readily seen that this type of flow pattern would produce a rapid exchange of heat and momentum between the "pole" and the "equator."

In addition to laboratory model experiments, studies of the general circulation can be carried out using high-speed electronic digital computers to solve the equations of atmospheric motion. In a typical computation it was assumed that initially the atmosphere was everywhere at rest relative to the earth and had a constant temperature. The radiation from the sun was then allowed to heat the atmosphere and the ground, and the effect of this heating on the motion of the air was computed. About thirty days after "the sun was turned on," the motions had settled down to a pattern similar to

**Figure 7.4** *Photograph of flow pattern in a "dishpan" laboratory model of the general circulation. [Courtesy of D. Fultz, University of Chicago Hydrodynamics Laboratory.]*

that observed on any day in the real atmosphere, with a Hadley circulation and easterlies at low latitudes, westerlies at higher latitudes and aloft, anticyclonic and cyclonic eddies in middle latitudes, and waves in the upper westerlies.

The theoretical computations and the laboratory model experiments both reproduce the principal features of the general circulation, showing that to transfer the heat energy in a rotating system the Hadley circulation must break down and be replaced at higher latitudes by a system of waves and vortices of the type described in Section 7.4.

## 7.6 The observed average-wind distribution

Figure 7.5 is a schematic representation of the average surface winds over the earth. In tropical regions, from about 30°S to 30°N latitude, the trade winds are present—northeasterly in the Northern Hemisphere and

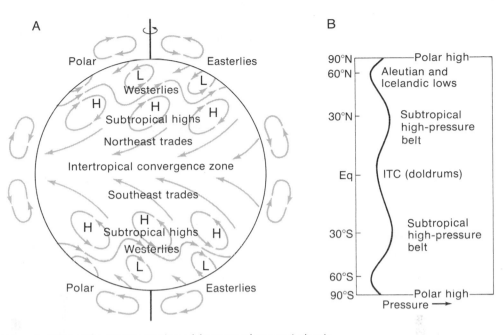

**Figure 7.5**  *A. Schematic representation of features of general circulation. B. Variation of average sea-level pressure with latitude.*

southeasterly in the Southern Hemisphere. Where the two systems come together there is a narrow zone of weak winds, called the doldrums by early mariners but now referred to as the *intertropical convergence zone (ITC)*. Located in the rising branches of the Hadley cells, this region is characterized by cloudiness and rain. Some of the photographs from ATS satellites have shown two bands of cloudiness in this area, suggesting that there may be a double structure, or two intertropical convergence zones, at times. The averaging of all the available cloud data from satellites shows that a single well-developed ITC is present most of the time.

In middle latitudes are the belts of westerlies, with waves and cyclonic vortices in them. Associated with the waves and vortices in the westerlies is highly variable weather, with rapidly changing temperatures and periods of clear skies followed by periods of precipitation. In winter the temperature may change as much as 20°C in a single day.

Between the westerlies and the trade winds are the subtropical high-pressure belts, called "Horse Latitudes" in the old days. These are regions of calms or light winds. Representing the poleward limits of the Hadley cells, where their descending branches occur, they are generally regions of fair weather and little precipitation. The deserts of the world are mostly at these latitudes.

In polar regions the winds on the average are easterly. The polar easterlies are not as persistent or steady as the trade winds, however. They are frequently interrupted by the passage of storms.

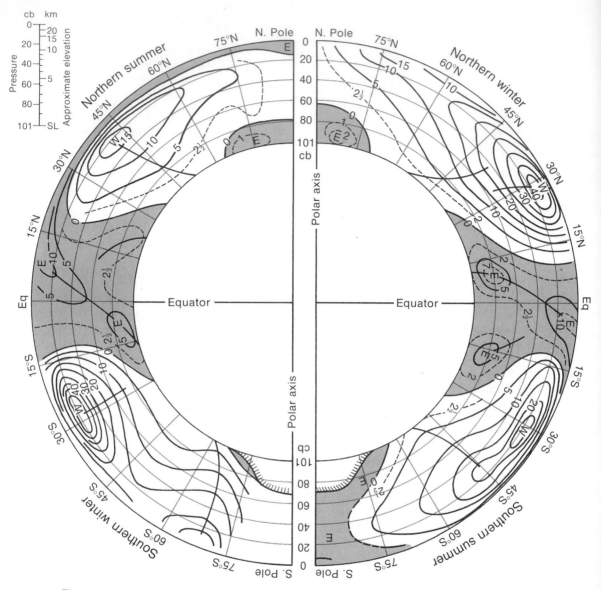

**Figure 7.6** *Mean zonal (west-east component) wind averaged around the earth at each latitude and height, in summer and in winter. Isotachs (lines of equal speed) are labeled in meters per second, with west winds positive. [After Y. Mintz, "The Observed Zonal Circulation of the Atmosphere," Bull. Am. Meteorol. Soc. 35 (5) (1954):209.]*

Figure 7.6 shows the observed zonal winds at all levels, averaged around the earth for summer and winter. The vertical scale is in pressure units, with the corresponding approximate elevation given in the upper left-hand corner.

It will be seen that the trade-wind easterlies extend a little farther poleward in the summer than the winter. The boundary between easterlies and westerlies slopes equatorward and the speed of the easterlies decreases with height in the troposphere. In the stratosphere there is a belt of stronger easterlies on the summer hemisphere side of the equator.

The strength and latitudinal extent of the westerlies increase with height in both hemispheres, both in winter and in summer, up to the tropopause. The highest velocities are at the poleward limits of the Hadley cells, where they might be expected from considerations of momentum conservation. In the winter there is a second, smaller maximum at high latitudes.

## 7.7 The observed mean pressure distribution

The wind distribution described in the preceding section has associated with it, as required by the geostrophic and gradient wind relationships, a corresponding distribution of pressure. The zonal average pressure distribution required for the surface wind distribution is shown in the diagram at the right of Figure 7.5. Aloft, the polar high pressure is replaced by low pressure in the Northern Hemisphere, and the subtropical high-pressure belt tilts equatorward.

If one considers the variation of heating with season one would expect a poleward shift of the pressure patterns from their average positions in summer and an equatorward shift in winter. The effect of continents and oceans, which respond differently to the sun's radiation, is to produce deviations from zonal symmetry. In summer the continents are warmer than the oceans, and the pressure tends to be higher over oceans and lower over continents, at least at low and middle latitudes. In winter the continents are cold and the pressure there tends to be higher than over the oceans at the ground and lower than over the oceans aloft.

These features may be seen in Figures 7.7 and 7.8, which show the pressure distribution in the Northern Hemisphere at sea level in January and July. At sea level in summer (Figure 7.8) the subtropical high-pressure centers are displaced poleward, to 35° or higher, over the ocean, but over land, particularly over Asia and northern Africa, there is low pressure instead of high at these latitudes. In winter, on the other hand, it is the low-pressure belt to the north of the westerlies that is interrupted by the effect of the continents, with deep centers of low pressure to the west of Iceland in the Atlantic Ocean and the Aleutian Islands in the Pacific, and high pressures over Asia and North America at the same latitude. The high-pressure belt at 30°N is more or less continuous around the earth.

The subtropical high-pressure centers over the oceans in summer and the subpolar low-pressure centers over the oceans in winter are semipermanent features. During the rest of the year they are present in nearly the same places most of the time, but they are usually less intense and their positions are more variable. They are known by the names of their locations: the Icelandic LOW, the Aleutian LOW, and the Bermuda or Azores

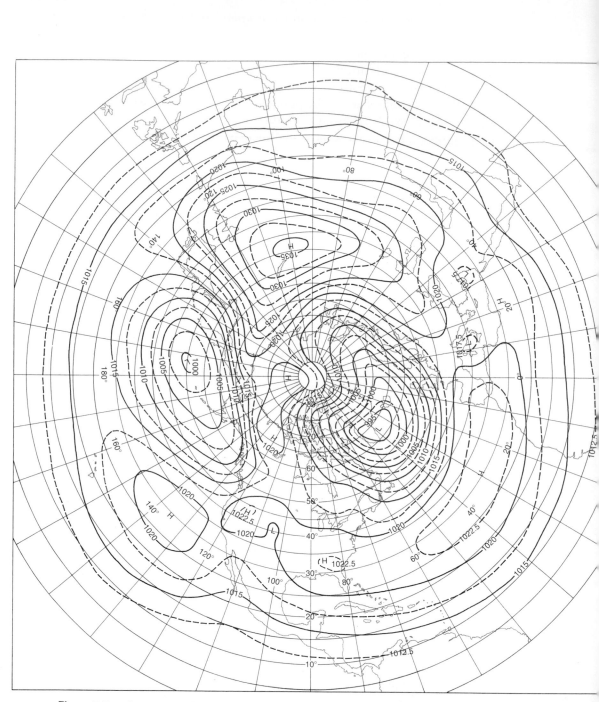

**Figure 7.7** *Average sea-level pressure in mb, Northern Hemisphere, January. [From U.S. Department of Commerce, Weather Bureau.]*

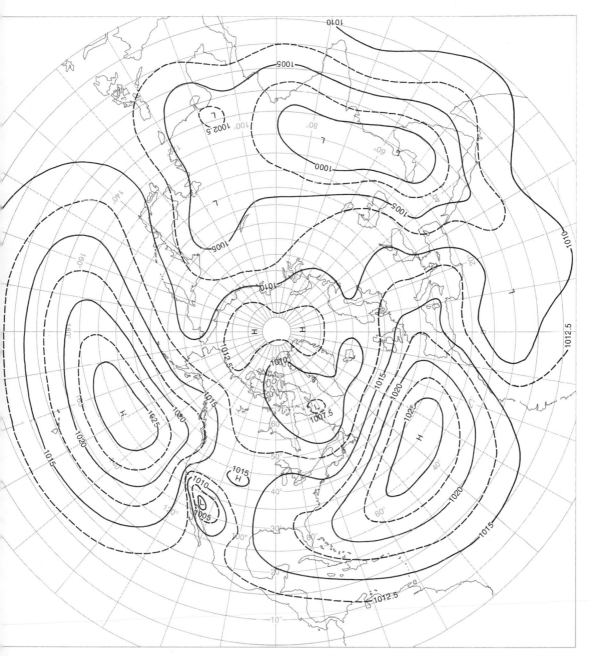

**Figure 7.8**   *Average sea-level pressure in mb, Northern Hemisphere, July. [From U.S. Department of Commerce, Weather Bureau.]*

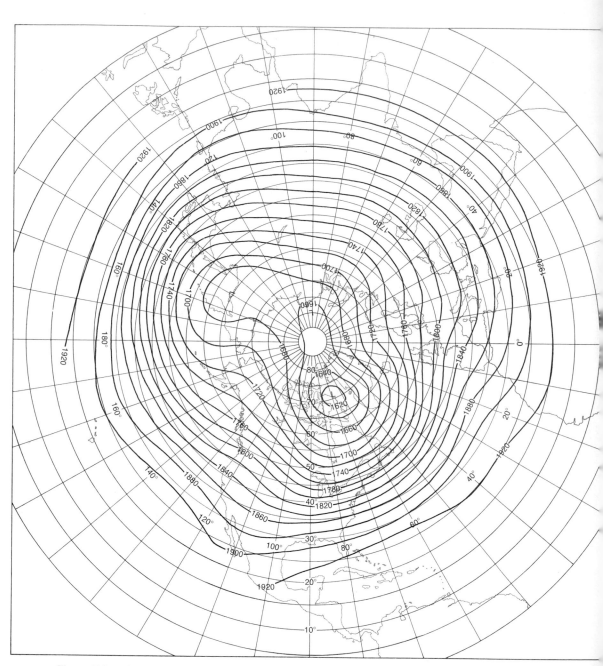

**Figure 7.9** *Average height, 500-mb surface, in tens of feet, Northern Hemisphere, January. [From U.S. Department of Commerce, Weather Bureau.]*

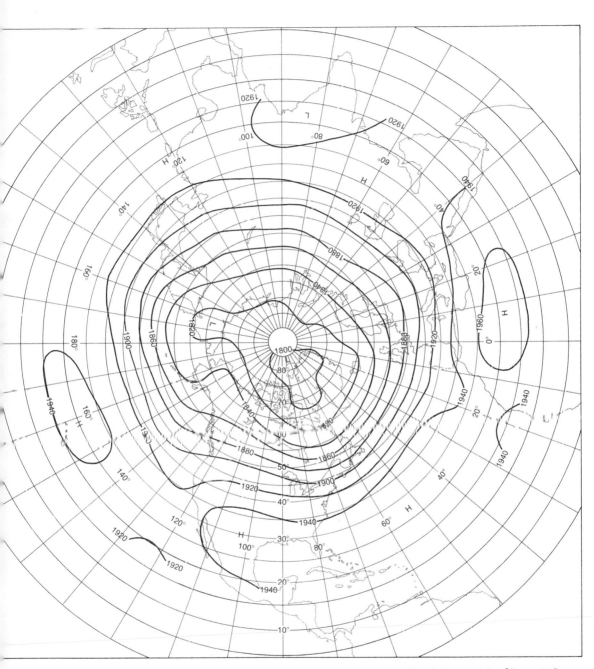

**Figure 7.10** *Average height, 500-mb surface, in tens of feet, Northern Hemisphere, July.* [*From U.S. Department of Commerce, Weather Bureau.*]

HIGH. (The subtropical HIGH of the North Pacific does not have a similar name.)

Aloft (Figures 7.9 and 7.10), the subtropical highs are found closer to the equator, at 25°N at 500 mb in July and still closer to the equator with no definite centers shown on the January map. The Icelandic and Aleutian lows are displaced westward to Greenland or eastern Canada and Siberia, and the high pressure near the North Pole is replaced by low pressure. The isobaric contours are much closer together in January than in July, corresponding to the greater temperature contrast between the polar and tropical regions in winter.

Because much less of the Southern Hemisphere is covered by continents the pressure pattern there is more nearly zonal, particularly in middle and high latitudes, and the seasonal variation is smaller. The subtropical high-pressure belt is broken up, with separate centers over the Atlantic, Pacific, and Indian oceans.

The intertropical convergence zone moves northward and southward with the seasons, from about 4°S latitude in January to 13°N in July.

### Questions, Problems, and Projects for Chapter 7

1   Discuss why there cannot be a simple Hadley circulation with easterlies at the ground at all latitudes from equator to pole.

2   Why is westerly momentum transported poleward by waves and vortices, even though the winds at low latitudes are easterlies? (*Hint*: Consider the conservation of the angular momentum of rings of air displaced poleward, as discussed in Section 7.3.)

3   a What would be the best route to sail from San Francisco to Hawaii? What is the best route to sail back?

   b Why does it usually take jet aircraft longer to fly from New York to San Francisco than from San Francisco to New York?

4   a Why are the low-pressure centers at about 60°N in Figure 7.7 over oceans and not over land?

   b Why are the subtropical high-pressure centers in Figure 7.8 over the oceans and not over continents?

5   a By looking at Figure 7.6, determine at which latitude the temperature changes with latitude most rapidly in the Northern Hemisphere (i) in winter and (ii) in summer.

   b Why does the boundary between the easterly trade winds and the westerlies tilt equatorward as one goes upward through the troposphere?

6   Why are the low-pressure troughs in Figure 7.9 over the continents and not over the oceans?

# EIGHT

## Water in the Atmosphere. Condensation and Precipitation

### 8.1 Humidity variables

In the discussion of the composition of the atmosphere in Chapter 2 water vapor was left for later consideration because of its variability. It is this variability, associated with the fact that it is constantly being added to the atmosphere by evaporation and removed by condensation and precipitation, that makes it such an important part of the air. The most conspicuous aspects of the weather—rain, snow, hail, fog, lightning, and so forth—result from the presence of water in the atmosphere.

The amount of water vapor in the air is termed *humidity*, but there are several ways in which humidity is expressed, such as absolute, relative, and specific humidity. The most commonly used of these, relative humidity, is perhaps the hardest to understand. Its significance will become clear after we have discussed some of the other measures of moisture in the atmosphere.

We can begin with the *absolute humidity*. This term is just another name for the water-vapor density, $\rho_v$; that is, the mass of water vapor per unit volume of air. In other words, if we were able to take all the molecules of water vapor in a unit volume and weigh them separately, we could find $\rho_v$ by dividing by the acceleration of gravity $g$. (Remember that the weight $W$ is the force acting on a mass $M$ due to the attraction by the earth: $W = Mg$.)

**Figure 8.1** *Snow crystals in the form of branched hexagonal plates or dendrites. [Courtesy of NOAA.]*

In an analogous fashion we can think of the pressure that is exerted by the molecules of water vapor as they bounce around. While they obviously collide with nitrogen and oxygen molecules, as well as with other water-vapor molecules, it turns out that the part of the total pressure due to their motions is the same as the pressure they would exert if no other gases were present. The part of the pressure they exert is called the water-vapor pressure, usually represented by the letter $e$, and the law of partial pressures just stated may be written

$$p = p_d + e \qquad (8.1)$$

where $p$ is the total pressure and $p_d$ the pressure due to the molecules composing dry air.

The vapor pressure is also a convenient measure of the humidity. It is related to the absolute humidity through the equation of state [see equation (4.1)]

$$e = \rho_v R_v T \qquad (8.2)$$

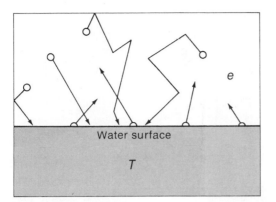

**Figure 8.2** *Schematic representation of water molecules leaving water surface in enclosed tank and of molecules returning. The saturation vapor pressure $e_s$, which the vapor pressure in the space above the surface reaches when the number of molecules returning is the same as that leaving, depends only on the water temperature T.*

where $R_v$ is the gas constant for water vapor. (Since vapors near the condensation point do not behave exactly like perfect gases, this equation is not quite correct, but it is a very good approximation.)

The vapor pressure and the absolute humidity of a parcel of air change when the pressure and temperature of the parcel change, even if no water vapor is added to or subtracted from it by evaporation, condensation, or mixing. It is desirable to have a humidity variable that remains constant when the temperature and pressure change. Such a variable is the *specific humidity*, $q$, or its equivalent, the *water-vapor mixing ratio*, $w$, defined as follows:

$$\text{specific humidity } q = \frac{\text{mass of water vapor}}{\text{mass of air}} = \frac{\rho_v}{\rho}$$

$$\text{mixing ratio } w = \frac{\text{mass of water vapor}}{\text{mass of dry air}} = \frac{\rho_v}{\rho_d} = \frac{q}{1-q}$$

Since the largest values of $q$ observed are just a few percent, the values of $q$ and $w$ differ at most by only this amount, and the two are used interchangeably.

To discuss the relative humidity we must explain the concept of saturation. This is because the ordinary statement that the relative humidity is the percentage of the water vapor that "the air will hold" is not correct. It has been demonstrated that in the absence of surfaces on which condensation can take place relative humidities of more than 400 percent can be produced. When we discuss the condensation process the reason for this will become clear.

**Figure 8.3** *Hygrothermograph. This instrument is a combination of a thermograph, for recording temperature, and a hygrograph, for recording the relative himidity. The humidity is sensed by strands of human hair (extending from top to bottom at the right end of the instrument). The hair gets longer when the relative humidity increases. A mechanical linkage transmits the change in length to the lower arm, at the end of which is a pen that draws a line on the chart as the drum is rotated by a clock. Similarly, the upper arm is controlled by a temperature sensor. In the instrument shown it is a curved tube filled with liquid; when the temperature rises, the liquid expands and straightens the tube. [Courtesy of Science Associates, Inc.]*

Let us consider a sealed tank partly filled with pure water (Figure 8.2). The upper part is initially assumed to contain dry air at the same temperature. Because of the thermal agitation of the water molecules some of them will escape from the surface (i.e., they will evaporate). As the evaporated molecules bounce around in the space above, some of them will bounce back to the water surface, and eventually the number returning to the surface will exactly equal the number leaving. When this state is reached the vapor above is said to be in *saturation equilibrium* with the water surface, and the pressure it exerts is called the *saturation vapor pressure $e_s$* with respect to a plane water surface at the temperature $T$ that it has attained.

The temperature is important, for the rate at which the molecules leave the surface, and thus the saturation vapor pressure, depends only on the temperature. It is immaterial whether the air pressure above the liquid water is high or low. The variation of saturation vapor pressure with temperature is shown in Figure 8.4. It has the value 6.1 mb at 0°C, and approximately doubles for every ten degrees. At 20°C, air thus "can hold" four times as much water vapor as at 0°C.

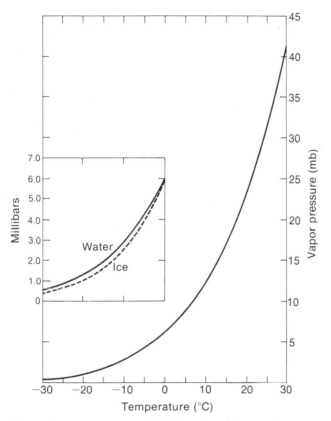

**Figure 8.4**  *Saturation vapor pressure (right-hand scale) over plane sur-*
*faces of pure water as a function of temperature.* Inset:
*Saturation vapor pressure (left-hand scale) over water and ice*
*at temperatures below 0°C.* [*From H. R. Byers,* Elements of
Cloud Physics *(Chicago: University of Chicago Press, 1965).*]

The *relative humidity*, $U$, is defined as the ratio of the observed vapor
pressure $e$ to the saturation vapor pressure $e_s(T)$ for the observed temperature. It is usually expressed in percent, so that

$$U - 100 \; e/e_s(T)$$

## 8.2  Condensation

When water condenses in the form of drops in the air it is obvious that it
must start with drops of very small radius, which then grow to larger drops.
The equilibrium vapor pressure $e_r$ over a small drop of radius $r$ is larger
than that over a plane water surface at the same temperature. In fact, it was
shown by William Thomson (Lord Kelvin) that $e_r$ is proportional to $1/r$. The
uppermost curve in Figure 8.5 shows how the equilibrium vapor pressure
over drops of radius $r$ varies with $r$. For very small drops, $e_r$ is many times
$e_s$. C. T. R. Wilson devised the cloud chamber, which subsequently was

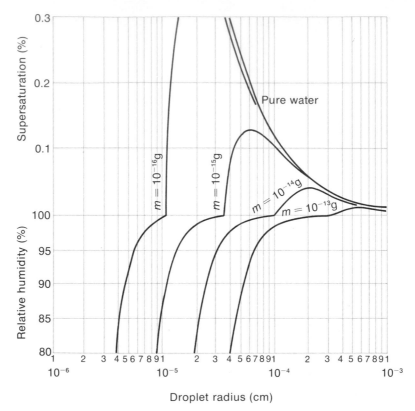

**Figure 8.5** *Relative humidity (for values less than 100 percent) and supersaturation (R.H. − 100) in equilibrium with drops of salt solution containing various masses* m *of common salt (NaCl). [From B. J. Mason,* Clouds, Rain, and Rainmaking. *(Cambridge: Cambridge University Press, 1962).]*

used so effectively in nuclear physics, in order to study this effect. Wilson found that when all dúst and ions were removed from air, supersaturations up to 700 percent occurred without condensation taking place. Under these circumstances, ionizing radiation left tracks of droplets condensing on the ions, showing the paths of the rays.

In the troposphere large supersaturations are never observed. In fact, the humidity rarely exceeds 100 percent by more than a few tenths. The reason for this is that the air always contains a large number of solid or liquid particles that can act as *condensation nuclei*. These particles favor formation of droplets at reasonable humidities in two ways. In the first place, they provide surfaces of larger radius of curvature so that the drops that form on them are in equilibrium at small supersaturations. For instance, from Figure 8.5 we see that if there are particles larger than 0.6 $\mu$m, drops can form on them if the supersaturation is 0.2 percent, and if there are nuclei larger than 1 $\mu$m, condensation on them will occur at 0.1 percent supersaturation. Actually, this is true if the nuclei are "wettable." Some substances are hygrophobic and resist condensation even if they are large enough.

In the second place, some substances "like" water and tend to absorb it even at humidities less than 100 percent. Such substances are called *hygroscopic*. By dissolving in the water, the hygroscopic nuclei reduce the equilibrium vapor pressure below that over pure water. The combined effects of size and dissolved salt are shown by the lower curves in Figure 8.5 for sodium chloride nuclei of several sizes. In each case the nucleus can start growing at humidities below 100 percent. In order to grow indefinitely the humidity must exceed 100 percent, but by amounts much smaller than would be required for a nonhygroscopic nucleus of the same size. For instance, a salt crystal of mass $10^{-15}$ g, with equivalent radius of 0.05 $\mu$m, which would require several percent supersaturation if it were not hygroscopic, requires only 0.13 percent supersaturation to grow indefinitely.

Hygroscopic nuclei are introduced into the atmosphere by evaporation of drops of sea spray and by combustion processes, both natural (forest fires) and artificial (domestic heating and industry). They vary in number with time and location, but in general it appears that there are enough large hygroscopic nuclei everywhere (say, more than 100/cm³) so that condensation will occur when the relative humidity exceeds 100 percent by a very small amount, perhaps less than 0.1 percent. For practical purposes it can be assumed that clouds will form whenever the relative humidity reaches 100 percent.

The relative humidity can change either by the increase of the vapor content (by evaporation) or by the decrease of the saturation vapor pressure (by cooling). If the relative humidity is raised above 100 percent by either of these processes, condensation will occur.

It may seem paradoxical on first sight, but there are circumstances in which evaporation produces supersaturation and condensation. Two cases are briefly described here.

1  If warm rain falls through cold air the rain will evaporate because the equilibrium vapor pressure for the warm water is much higher than the vapor pressure of the cold air even when the air is "saturated." As soon as the vapor is cooled to the temperature of the air, however, it will be supersaturated and will condense on the nuclei in the air. This type of condensation occurs sometimes in the cold air ahead of a warm front, when it forms *warm frontal fog*.

2  If cold air flows over warm water, for instance polar air passing onto an unfrozen lake in winter or air moving across a heated swimming pool at night, the vapor evaporating from the warm water will be cooled by mixing with the air, producing supersaturation and condensation on nuclei. This type of condensation is called *steam fog* or *arctic sea smoke*.

The processes by which the relative humidity can increase to 100 percent by the lowering of temperature are cooling by radiation, cooling by conduction and turbulent transfer of heat by eddies, and adiabatic cooling by decrease of pressure. Cooling by radiation and cooling by conduction and turbulent transfer of heat by eddies ordinarily occur only close to the ground and cause fog. Like cases (1) and (2) just described, they cannot cause deep clouds and large-scale precipitation. At most they cause drizzle, which is very light precipitation composed of very small drops.

Fog due to radiational cooling is called *radiation fog*. It occurs on clear nights with high moisture content in a shallow layer near the ground and very low humidities aloft. Under these circumstances the ground cools rapidly, since the radiation from the ground is only slightly offset by the radiation downward from the air above because of the low total water-vapor content of the air column. Once the ground has become sufficiently cooler than the moist layer immediately above it, the moist layer absorbs less radiation than it emits and cools, ultimately becoming saturated and producing fog. In light winds the radiative cooling of the air is augmented by turbulent transfer of heat from the moist layer to the colder ground. If the winds are not light, however, the turbulence will produce mixing of the surface air with the warmer, drier air aloft and prevent the formation of radiation fog. The requirement for clear nights, light winds, and low humidities aloft is met principally in stagnating high-pressure areas, particularly in the fall and winter. In the winter the fog, once formed, may persist throughout the day because the amount of solar radiation, reduced by reflection from the fog top, may not be sufficient to heat the air enough to dissipate it. Thus in the valleys of the Great Basin and the interior of California, fog may be present for days and even weeks at a time. In the warmer part of the year, with larger amounts of insolation, radiation fog normally is dissipated within an hour or two after sunrise.

The formation of radiation fog takes place in two steps: first the cooling of the ground by radiation to "space," and then the cooling of the moist air. The same effect as the first step is produced in another way in *advection fog*. In advection fog warm moist air moves, or is "advected," over a cooler surface; when a sufficiently large temperature difference is produced, the transfer of heat from the warm moist air to the cold ground lowers the air temperature and produces fog. For example, advection fog occurs when air from the Gulf of Mexico moves over the southern United States in winter, or when air is carried northward by south winds from the Gulf Stream area of the Atlantic Ocean to the frigid waters of the Grand Banks of Newfoundland. Because the temperature contrast between ground and air tends to be greater when the winds are stronger, advection fog can occur in moderately strong winds in spite of the tendency of turbulent transport to dissipate it.

While radiative cooling ordinarily produces condensation only close to the ground, decrease in pressure due to rising motions can cause cooling through deep layers at various heights. The adiabatic cooling associated with rising air currents is responsible for the formation of almost all clouds and precipitation.

In addition to air moving upward, air moving horizontally across isobars, which occurs principally in the friction layer, undergoes small decreases of pressure and associated adiabatic cooling. The cooling that results from cross-isobaric flow contributes to the formation of fog ahead of warm fronts.

## 8.3    The saturation adiabatic process

When a substance changes phase, for instance, from liquid to solid or from liquid to gas, a certain amount of heat energy is either released or absorbed. To see that this occurs, consider what happens when a kettle containing water is placed on the burner of a stove. At first the temperature of the water rises as the heat is added, but once it starts boiling the water temperature remains constant (100°C at standard sea-level pressure). The heat continues to be added, but is used entirely to change the liquid water to water vapor (steam). The amount of heat required to convert liquid water to vapor at 100°C is 540 calories per gram, or $2.26 \cdot 10^6$ joules per kilogram. Because this heat does not show up as a change in temperature it is called *latent heat*. When water evaporates at temperatures lower than the boiling point, the amount of latent heat is larger. At 0°C it is 596 calories per gram, or $2.500 \cdot 10^6$ joules per kilogram. The same amount of latent heat is given off during condensation as is consumed in evaporation at the same temperature.

When a solid changes to liquid a similar though smaller amount of latent heat is taken up. At 0°C the latent heat of melting of ice (or freezing of water) is 80 calories per gram, or $0.334 \cdot 10^6$ joules per kilogram. If the solid changes directly to vapor without melting first, as when snow evaporates, the latent heat is the sum of the latent heat of melting and the latent heat of vaporization. It is called *the latent heat of sublimation*, totaling 676 calories per gram or $2.834 \cdot 10^6$ joules per kilogram at 0°C.

Let us now consider a rising air parcel that has cooled adiabatically to the point where it is slightly supersaturated and condensation on the nuclei in it begins. As the water condenses on the nuclei, latent heat of condensation is released and tends to offset the cooling due to adiabatic expansion. The rate of decrease of temperature of the rising cloudy air is thus less than the adiabatic process rate in the absence of condensation. To distinguish the two we refer to the latter as the *unsaturated* or *"dry" adiabatic rate of cooling*, and to the process rate with condensation as the *saturation adiabatic rate of cooling*. Unlike the unsaturated process rate $\Gamma$, which is constant, the saturation adiabatic rate $\Gamma_s$ (sometimes called "wet" adiabatic rate) varies with the temperature and pressure. At high temperatures it is less than 0.5°C/100 m, but at very low temperatures it approaches 0.98°C/100 m, the value of $\Gamma$. The values of $\Gamma_s$ for several pressures and temperatures are given in Table 8.1.

Saturation adiabatic processes can be represented by lines on thermodynamic diagrams similar to the unsaturation process curves. Since the saturation adiabatic rate of cooling is smaller, the saturation adiabatic

**Table 8.1** Saturation Adiabatic Process Rate (°C/100 m)

| Temperature (°C) | Pressure (mb) | | | | |
|---|---|---|---|---|---|
| | 1000 | 850 | 700 | 500 | 300 |
| 40 | 0.30 | 0.29 | 0.27 | | |
| 20 | 0.43 | 0.40 | 0.37 | 0.32 | |
| 0 | 0.65 | 0.61 | 0.57 | 0.51 | 0.41 |
| −20 | 0.86 | 0.84 | 0.81 | 0.76 | 0.68 |
| −40 | 0.95 | 0.95 | 0.94 | 0.93 | 0.90 |

curves, called *saturation adiabats* or *wet adiabats*, lean less toward lower temperatures. As they go upward toward lower pressures and temperatures, they become more nearly parallel to the dry adiabats.

In Figure 8.6 some saturation adiabats are shown, in addition to the other lines described in Section 4.4, on another type of thermodynamic diagram called the *Skew T-Log P Diagram*. This diagram is similar to the one previously described, except that the isotherms slope upward to the right, so that the angle between the isotherms and the dry adiabats is about 90°. Since most actual sounding curves and process curves lie between the isotherms and the dry adiabats, it is advantageous to use a diagram on which this angle is large. The isobars, as before, are horizontal lines and the dry adiabats slope upward to the left. The saturation adiabats are the curves sloping slightly to the left near the bottom of the diagram and curving farther to the left as they go up.

In the same way that the potential temperature serves to identify the dry adiabats, the saturation adiabats are labeled by the temperature at which they intersect the 1000-mb isobar. This temperature is called the *wet-bulb potential temperature* of any saturated air parcel whose representative point falls on the saturation adiabat.

Suppose that a parcel having a temperature of 15°C, pressure of 1013 mb, and relative humidity of 58 percent moves upward. The representative point for its initial condition is point $A$ in Figure 8.6. As it rises the representative point will follow the dry adiabatic process curve $AB$ until it is saturated, and then the saturation adiabat $BC$.

How do we know it will become saturated at $B$? For this purpose meteorological thermodynamic diagrams have still another set of lines on them, the *vapor lines*, drawn for constant values of saturation mixing ratio $w_s$. These are shown in Figure 8.6 by dotted lines sloping upward to the right, forming small angles with the isotherms. They are labeled in parts per thousand, frequently expressed as "grams per kilogram". Thus saturated air at point $A$ would have a mixing ratio of $10.5 \cdot 10^{-3}$. Since the relative humidity is 58 percent the actual mixing ratio is approximately $0.58 \cdot 10.5 \cdot 10^{-3} = 6.1 \cdot 10^{-3}$. When the air has risen adiabatically until its pressure and temperature have the values at which $w_s = 6.1 \cdot 10^{-3}$, which occurs at point $B$, it is saturated and from there on it will follow the saturation adiabat.

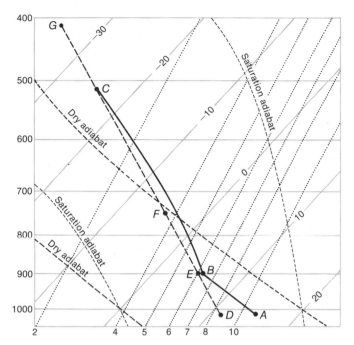

**Figure 8.6** *Thermodynamic diagram (Skew T-Log P Diagram), showing saturation adiabats (curved dashed lines) and saturation mixing ratio lines (dotted lines). The process that a parcel with an initial temperature of 15°C, a pressure of 1013 mb, and a mixing ratio of $6.1 \cdot 10^{-3}$ would experience if lifted adiabatically is represented by curve ABC.*

The pressure and temperature at which the saturation is reached in this fashion are called the *lifting condensation pressure* and the *lifting condensation temperature* of the air represented by point $A$, and the corresponding height is the *lifting condensation level*. These values are determined by the initial properties of the air parcel. In turn, they determine the saturation adiabat that the parcel at $A$ would follow if it rises sufficiently. Thus the wet-bulb potential temperature it would have once it became saturated is determined by the initial properties of the air parcel, and it is considered to have that wet-bulb potential temperature whether or not it actually rises and reaches saturation. We can see that the wet-bulb potential temperature of an air parcel is conserved (does not change) for any adiabatic process, unsaturated or saturated. The potential temperature, on the other hand, changes during saturation adiabatic processes, although it is constant for unsaturated adiabatic processes.

Let us now consider what happens if, instead of a single parcel rising, there is turbulent mixing of deeper and deeper portions of the layer of air next to the ground. The mixed layer attains a constant potential temperature and a constant mixing ratio, which are the averages of the original values of these quantities through the layer. By this mixing process it is possible that the air at the top of the mixed layer will become saturated. Only in the situation in which the potential temperature and mixing ratio were initially constant through the layer would the level at which saturation begins be the same as in the previous case. The pressure and temperature at which saturation is reached by turbulent stirring are called the *mixing condensation pressure* and the *mixing condensation temperature*.

## 8.4 Stability of cloudy air

Since the rate of change of temperature when cloudy air rises is different from that of unsaturated air, the criteria for stability of a saturated air column are different from those for unsaturated air. The same type of reasoning as that used in discussing Figure 4.6 leads us to the following criteria for saturated or cloudy air.

If $\gamma < \Gamma_s$, the air column is in stable equilibrium.

If $\gamma = \Gamma_s$, the air column is in neutral equilibrium.

If $\gamma > \Gamma_s$, the air column is in unstable equilibrium.

Note that since $\Gamma_s < \Gamma$, these criteria lead to the possibility that an air column would be in stable equilibrium if it were unsaturated and in unstable equilibrium if it were saturated. This state is called *conditional instability;* it occurs if $\gamma$ has a value between $\Gamma_s$ and $\Gamma$, that is, if $\Gamma > \gamma > \Gamma_s$.

Because $\Gamma_s$ varies it is not easy to apply these criteria directly. The convenient procedure is to plot the sounding curve on a thermodynamic diagram and compare its slope for the saturated portions of the air column represented with the saturation adiabats.

Suppose that the observed sounding is represented by the dashed curve *DEFG* in Figure 8.6. By comparing the sounding curve with the saturation adiabats, we see that below point *F*, $\gamma > \Gamma_s$ and above point *F*, $\gamma < \Gamma_s$. If the air at any place in the column represented by *DF* is saturated, it is unstable. The air at any position in the column represented by *FG* is stable, whether or not it is saturated.

Suppose, further, that the air represented by point *D* has a mixing ratio of $6.1 \cdot 10^{-3}$ and is heated from 11°C to 15°C (by contact with the heated ground). It then will be represented by point *A*, and a slight upward push would cause it to rise and be continuously accelerated because it would be warmer than its surroundings. Its representative point would follow the dry adiabat *AB* as before. At the elevation represented by point *B*, condensation would begin, and the representative point of the air parcel would follow the saturation adiabat *BC*. When it reached *C* it would be in equilibrium with the surroundings, and further ascent would subject it to a downward buoyant force. By this process we would expect a cloud to be formed that would extend from about 900 mb (1 km) to about 500 mb (5.5 km). (We shall see in the next chapter that the process of entrainment influences the growth of clouds produced in this fashion.)

At point *C* the mixing ratio of the air parcel would be the saturation mixing ratio there, $1.0 \cdot 10^{-3}$. The difference between this value and the original value would be the amount of liquid water condensed, $5.1 \cdot 10^{-3}$. At the level represented by $C$ the air density is about 0.7 kg/m³, so that the liquid content of the cloud that formed, assuming none fell out of the parcel, would be 3.6 g/m³ or $3.6 \cdot 10^{-6}$ g/cm³. By assuming a reasonable value for the number of nuclei on which the condensation occurs, we can estimate the average size of the drops in the cloud. For instance, if 100 nuclei per cm³ were activated, each drop would have a mass $3.6 \cdot 10^{-6} \div 100 = 3.6 \cdot 10^{-8}$ g. Since the density of water is unity, the average volume of each drop, $4/3 \pi r^3$, is equal to $3.6 \cdot 10^{-8}$. From this we find that $r$, the average radius, is equal to 20.5 $\mu$m. As we shall see in the next section, a drop of this size is too small to fall as precipitation. Other processes are required to produce precipitation-sized particles.

## 8.5  Clouds and precipitation

One might wonder why all clouds do not precipitate. The water drops (or ice crystals) are heavier than the air and should fall. The explanation lies in the fact that the drops in fog and clouds are so small that they fall very slowly through the air. If the air is moving upward (and most clouds form because of upward motion), the air movement will more than compensate for the downward motion of the drops. Only when the drops have grown large enough to have a large velocity of fall will they overcome the upward air motion and reach the ground as rain.

Again, this difference in fall velocity between small and large drops may be puzzling. Galileo demonstrated that gravity acts to produce the same acceleration on all bodies, light or heavy. In a vacuum, small and large drops would behave alike. But the resistance of the air to the motion depends on the size, as well as the speed, of the falling body in such a way that the resistance increases more rapidly than the force of gravity as the size and weight get larger. The result is that there exists a speed, called the *terminal velocity,* at which the force due to air resistance equals the force of gravity. When the speed of the falling body reaches this value, the forces are in balance, and it is no longer accelerated but falls thereafter at a constant speed. In general, the larger and heavier the body, the greater its terminal velocity.

It is of interest to note that this relationship has its implications in the animal world. When small animals such as ants or even mice fall from great heights they are unhurt and scurry away, because their terminal velocities are sufficiently small. Larger animals, especially humans, have such large terminal velocities that they are injured or killed by the impact of a fall.

For water drops the terminal velocity depends only on the radius (or equivalent radius for large drops that are distorted from spherical shape when falling). It varies directly as the square of the radius for small drops,

**Table 8.2** Terminal Velocities of Water Drops in Air

| Radius ($\mu$m) | Velocity (cm s$^{-1}$) | Radius (mm) | Velocity (m s$^{-1}$) |
|---|---|---|---|
| 1 | 0.013 | 0.25 | 2.06 |
| 5 | 0.32 | 0.5 | 4.03 |
| 10 | 1.3 | 1 | 6.49 |
| 20 | 5.0 | 1.5 | 8.06 |
| 50 | 27 | 2 | 8.83 |
| 100 | 72 | 2.5 | 9.09 |

(Note the change in units between the left and right pairs of columns.)

and approximately as the first power of the radius for larger drops. Table 8.2 gives some representative terminal velocities.

We note that a drop having a radius of 5 $\mu$m would fall (in the absence of air motion) about 12 meters in an hour. It would take more than 24 hours for it to fall from a cloud base at 300 m (1000 ft) to the ground. In a much shorter time it would evaporate if the air below the cloud were not saturated. On the other hand a drop having a radius of 1 mm would fall 300 meters in less than a minute, and it would take large updrafts to offset its falling.

We see that if clouds consist entirely of small drops, with radii of less than 20 $\mu$m, very slight upward motion of the air will prevent precipitation, but if they contain drops larger than one- or two-tenths of a millimeter, precipitation is likely.

Observations show that clouds do indeed consist of small drops. For all types of liquid clouds the most frequent drop radius is between 5 and 10 $\mu$m. There are smaller and larger drops but few, if any, are larger than 20 $\mu$m. The number of drops ranges between 30 and 300 per cubic centimeter, with 100 per cubic centimeter being a representative value.

It is the large number of drops that leads to their small size by the condensation process. For instance, even if all the water vapor were condensed out of fairly moist air, say with a mixing ratio of $10^{-2}$, the average radius would be 30 $\mu$m, if 100 drops per cubic centimeter were formed. In actuality, only a small fraction, perhaps a few percent, of the water vapor in the air condenses, and, consequently, the drops are much smaller than that. The reason why condensation does not lead to drops large enough to precipitate is that condensation nuclei that can be activated at small supersaturations are always present in large numbers.

Another factor is the rate at which condensation can take place. While the initial condensation on nuclei forms a cloud quite quickly once saturation is attained, computations show that the rate of growth decreases as the size increases, and it would take several hours to form precipitation-sized drops by condensation even if there were few enough condensation nuclei present. Since showers sometimes take place within one-half hour of the first formation of a cloud, it is clear that other processes must be responsible for the growth of cloud droplets to rain drops.

The amount of growth required may be seen by considering how many typical cloud drops are required to make a typical rain drop. We have seen that cloud drops are mostly in the 5–10 $\mu$m radius range, while rain drops are of the order of 1 mm (1,000 $\mu$m). Since the mass or volume varies with the cube of the radius, we see that it would take about *one million* cloud drops to form a single rain drop.

The two processes that have been found to be important in the growth of precipitation particles are—

1  the collision-coalescence process (the "warm cloud" process);
2  the three-phase process (the Bergeron process).

The collision-coalescence process is the easier to understand. Since clouds consists of drops of various sizes, the larger drops will be falling relative to the air faster than the smaller ones. If a smaller drop is directly below a large one, the large one will catch up with the smaller drop and should be expected to collide and combine with it. After that, being still larger, it will fall still faster, collecting more and more small drops as it falls, and eventually (after a million collisions) it will come out of the cloud as rain.

Even this process, however, is more complicated than it seems on the surface. For as the large drop falls through the air, it pushes the air ahead of it out of the way. This air moves aside and tends to drag the small drop aside with it. Thus even though the small drop is initially in the path of the large one, it may be carried around it by the air rather than collide with it.

The situation is illustrated in Figure 8.7. Here the path of the small drop is pictured approaching the large drop, the way it would appear if one were moving with the large drop. The paths shown are those for which the droplets would just touch. Should the small drop start farther away than $R$ from the axis of fall of the large drop, the drops would miss each other. The ratio $R/A$, where $A$ is the radius of the large drop, is called the linear collision efficiency $y_c$. It turns out that $y_c$ depends on $A$ and on the ratio $p = a/A$, where $a$ is the radius of the small droplet. Figure 8.8 shows the computed values of $y_c$ for various values of $A$.

The fraction of small drops collected, the collision efficiency, is proportional to $y_c^2$. We see from Figure 8.8 that when $A$ is less than 20 $\mu$m, the collision efficiency is extremely small. Thus it would appear that drops larger than 20 $\mu$m in radius are needed to initiate precipitation. Since most clouds formed by condensation do not contain drops of this size, it is necessary to call on the second process, the three-phase process, to explain most occurrences of rain. It is called the three-phase process because it requires the presence of ice crystals in addition to liquid drops and water

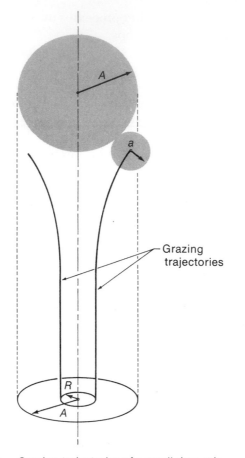

**Figure 8.7** *Grazing trajectories of a small drop relative to a large drop, illustrating the concept of collision efficiency. The linear collision efficiency $y_c$ is defined as the ratio R/A.*

vapor. As such, it can occur only at temperatures below 0°C. In some cases precipitation starts from clouds that are warmer than 0°C throughout their extent. These exceptional cases must have fewer condensation nuclei or some very large condensation nuclei (giant salt nuclei) to lead to the formation of drops larger than 20 $\mu$m by condensation.

The three-phase process requires the simultaneous presence of water drops and ice crystals at temperatures below "freezing." The occurrence of supercooled water drops is quite common. In fact, it is unusual for clouds to consist of ice rather than liquid at temperatures between 0°C and −10°C. Between −10°C and −20°C liquid and ice-crystal clouds occur, and below −20°C clouds are generally composed of ice particles.

The reason why liquid clouds occur at "sub-freezing" temperatures is similar to the reason why condensation in dust-free air requires great super-saturations. In the absence of appropriate nuclei, ice will not form at temperatures above about −40°C. Under ordinary conditions in lakes and

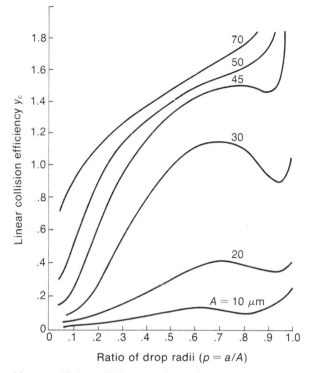

**Figure 8.8**  *Linear collision efficiency* $y_c$ *for various values* A *of the radius of a large water drop falling relative to a small cloud drop of radius* a, *as a function of the size ratio* p = a/A.

streams or in the ice-cube trays in a refrigerator, freezing nuclei are present or the freezing is nucleated at the solid shores of the lake or at the walls of the tray. In a drop floating in the air, nuclei that are effective near 0°C are less likely to be present.

Ice nuclei are of two types: freezing nuclei, which initiate the change from liquid water to ice, and sublimation nuclei, which facilitate the change from vapor to solid phase. In both cases the nuclei are temperature dependent. The most common natural nucleus, kaolinite (a form of clay), is effective at −9°C, and this explains why clouds frequently remain liquid at least down to this temperature.

The number of ice nuclei present in the atmosphere that are effective at temperatures higher than −20°C tends to be much smaller than the number of condensation nuclei effective at small supersaturations. Whereas the latter is of the order of 100 per cubic centimeter, the number of ice nuclei is typically one per liter. Thus even at temperatures at which ice nuclei are effective there would initially be only one ice crystal amid thousands or hundreds of thousands of liquid drops.

At temperatures below 0°C the equilibrium vapor pressure over ice is lower than that over water. In Figure 8.4, the insert shows the difference.

Curve $a$ is for liquid water and curve $b$ is for ice. If air at $-10°C$ is saturated with respect to liquid water it will be supersaturated with respect to ice.

Let us now visualize the situation in which an ice crystal forms in the midst of a myriad of liquid drops. The air initially is saturated with respect to the liquid drops and thus considerably supersaturated with respect to the ice crystal. Therefore more vapor molecules will condense on the crystal than will evaporate from it and the crystal will grow. As the vapor condenses, the vapor pressure is lowered below the saturation with respect to the drops, and the drops start evaporating. This process, by which the vapor is evaporated from the drops and condensed on the ice crystal, is called the three-phase process, or, after the Swedish meteorologist who first pointed out its importance, the Bergeron process. As this process continues, the water is rapidly transferred to the ice crystal from the many surrounding drops. The crystal grows much larger than the drops and starts falling relative to them. At this time, in addition to growing by the three-phase process, it will grow by collision and coalescence, once it has grown larger than the minimum size for nonzero collision efficiency.

The fact that most occurrences of precipitation require the three-phase process to initiate them, and that this usually occurs naturally only when the cloud tops are high enough for their temperatures to be below $-10°C$ or $-15°C$, has led to attempts to stimulate precipitation and augment its amount artificially. Vincent Schaefer discovered that ice nucleation could be initiated by dropping solid carbon dioxide (dry ice) pellets through a supercooled cloud, and Bernard Vonnegut found that silver iodide particles were effective ice nuclei at temperatures as high as $-4°C$. Many cloud-seeding experiments have been carried out, both with dry ice dropped from airplanes and with silver iodide smoke generators operated at the ground or on planes. In general, it may be said that while some of these experiments appear to have increased the amount of precipitation, we do not yet have adequate knowledge to discriminate between the conditions under which seeding will augment precipitation and those under which seeding will suppress cloud development and decrease precipitation. (We know that circumstances of both types can exist.) Weather modification is discussed in more detail in Chapter 15.

## 8.6 Forms of precipitation

Precipitation formed by the three-phase process is initially in the form of ice. When the temperature at all levels right down to the ground is below $0°C$, or at the highest only slightly above $0°C$, it remains frozen, but if the temperature of the lowest layers is more than a little above $0°C$, the particles usually melt and form raindrops. An exception is hail, which consists of pellets of ice that fall so rapidly that they are still solid when they reach the ground, even at fairly high temperatures.

**Figure 8.9** *Instruments for measuring precipitation. A. The standard rain gauge used by the U. S. National Weather Service for periodic observations—for instance, the total daily rainfall. Precipitation that has fallen into the gauge is transferred to a measuring tube with a cross-sectional area one-tenth that of the collector, so that the depth is magnified ten times for measurement. When continuous records are required, a recording rain gauge is used, of which the weighing rain gauge shown in part B is one example. The precipitation drains from the collector into a container that rests on a scale. The weight of the precipitation raises the pen arm, which records the amount on the chart on the clock-driven drum. [Courtesy of Science Associates, Inc.]*

Precipitation includes all forms of water particles that fall to the ground. The following are some of the forms it takes.

*Drizzle* is a fine mist of tiny liquid drops, smaller than 0.5 mm in diameter, which fall from stratus clouds, usually as a result of the warm cloud process. The amount of precipitation resulting from drizzle is very small, accumulating at most at the rate of 1 mm/hr.

*Rain* is the name given to all liquid precipitation other than drizzle. By definition the smallest raindrops are 0.5 mm in diameter. There is a natural limit to the largest size. Drops larger than 5 mm in diameter are unstable and break up into smaller ones as they fall. As shown in Table 8.2, the terminal velocity of drops having a diameter of 5 mm is 9 m s$^{-1}$. In clouds in which the updrafts exceed this speed, no drops can fall toward the ground. Steady, continuous rain falls from layer clouds that are associated with fronts and cyclones. Showers, which consist of rain with rapid changes in intensity and sudden starts and stops, fall from convective clouds. Rain is characterized as light if it falls at a rate less than 0.5 mm/hr, moderate from 0.5 to 4 mm/hr, and heavy if greater than 4 mm/hr.

*Snow* is the general designation for precipitation of opaque or semi-opaque ice particles in the form of individual crystals, small pellets, or flakes formed by aggregation of crystals. While the other forms are not uncommon, in most snowstorms the precipitation is in the form of flakes. Snowflakes range in size from a few millimeters to several centimeters. At low temperatures they tend to be small; at temperatures near 0°C they may be very large.

*Ice crystals* is the designation for precipitation in the form of single, individual crystals. They may fall from stratus clouds, or they may appear to come from a clear sky, falling as fast as they form in the rapidly cooling air. In the latter case, the designation "diamond dust" is used because the crystals sparkle in the light.

*Snow pellets*, also called soft hail or graupel, are white opaque spherical particles having a snowlike character. They are crunchy and easily crushed, in contrast to ice pellets and hail, which are hard ice. They usually occur in snow showers falling from convective clouds, particularly in mountainous areas.

Snowfall is measured both in the actual depth of snow accumulating and in equivalent liquid content. Depending on the type of snow, usually determined by the temperature, the liquid content of snow may vary greatly. Light fluffy snow may be as much as twenty times as deep as the equivalent liquid, while heavy "wet" snow may be as little as six times its depth when melted. A rough average value is that snow is about ten times the depth of the equivalent rain. The amount of precipitation given in weather reports is the equivalent melted depth.

*Ice pellets* are small transparent or translucent quasi-spherical particles of ice. They are of two types. One, which is also called *sleet*, is formed by the freezing of small raindrops as they fall. In the situation in which sleet occurs, a layer of air having temperatures above 0°C overlies a colder layer near the ground. Precipitation, formed at greater heights in the form of ice crystals, melts in falling through the warm layer, and then the drops formed thereby freeze while falling through the cold air below it. A "sandwich" of warm air like this is present sometimes at warm fronts. More serious from

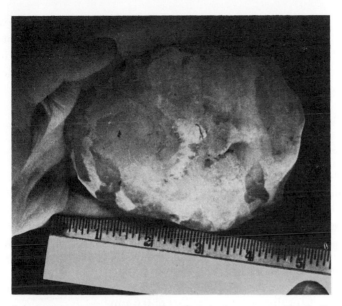

**Figure 8.10**  *A hailstone measuring more than four inches across. This hailstone fell during a thunderstorm at Weatherford, Oklahoma, July 1, 1940. [Courtesy of NOAA.]*

the standpoint of damaging weather are the situations that cause *ice storms.* In these situations the vertical temperature distribution is similar to those that produce sleet, but the subfreezing air at the ground is not deep enough or the raindrops are too large for the drops to freeze as they fall. In that case the rain freezes on everything it strikes on the ground. The resulting *glaze* on roads and walks creates a serious hazard to travel, and tree branches and utility wires sag and break because of the weight of the accumulated ice.

The other form of ice pellets is called *small hail.* It is formed when ice crystals or small snow pellets fall through supercooled cloud drops sufficiently rapidly to accumulate a coat of clear ice.

*Hail* is precipitation in the form of quasi-spherical or irregular lumps of ice. In size hailstones range from 5 mm — smaller particles are by definition termed ice pellets or snow pellets — to 10 cm or more in diameter, the most frequent size being about 1 cm. Hail forms in regions of very strong updrafts and high content of supercooled liquid in cumulonimbus clouds. The hailstones are composed of alternate layers of clear ice and opaque ice, the latter rendered opaque by the presence of numerous small air bubbles. The layered structure is attributed to differences in the rate of accumulation and freezing of the supercooled water. When the hailstone falls through air with high liquid content the water accumulates faster than it can freeze, and a coat of liquid forms, which becomes a layer of clear ice when it freezes. When the hailstone falls through air with smaller and less numerous cloud drops, they freeze immediately on impact, trapping bubbles of air as they freeze.

When storms with moderate to large hailstones occur they cause damage to crops, to structures, and to vehicles. In addition to attempts at weather modification aiming to increase precipitation where more rain is needed, experiments have been carried out to try to reduce or eliminate crop-damaging hail.

### Questions, Problems and Projects for Chapter 8

1   What is meant by saturated air? Why doesn't condensation occur when the relative humidity is slightly in excess of 100 percent in the absence of particles suspended in the air?

2   Suppose that the relative humidity is 100 percent and the temperature is 10°C at sunrise, and that during the day the air is heated but its water vapor content does not change. What will the relative humidity be when the temperature reaches 20°C? (*Hint*: Can you get the data you need from Figure 8.4?)

3   a From Figure 8.5, find what would be the relative humidity in equilibrium with a pure water drop having a radius of 0.6 $\mu$m.

   b What would happen if the ambient humidity had this value and the drop grew slightly larger, say, to 0.61 $\mu$m? What would happen if it became slightly smaller?

   c What is the relative humidity needed to enable a condensation nucleus containing $10^{-15}$ g of NaCl to grow indefinitely?

4   Why does the vapor pressure change when an unsaturated air parcel rises adiabatically without mixing with its surroundings? Does its mixing ratio change too? What about its relative humidity?

5   Why is the saturated adiabatic process rate of cooling smaller than the dry (unsaturated) rate? Why is the difference between them smaller at lower temperatures?

6   In Figure 4.7 a typical vertical temperature sounding for Santa Monica is represented. Suppose you heated a parcel of air at the ground to 25°C, and then released it. To what height would it rise if it moved adiabatically without mixing until it came into equilibrium with the surrounding air? Suppose that as you heated it you added moisture until it became saturated and then released it? To what height would it rise in that case? (*Hint*: An approximate value for the saturation adiabatic process rate can be obtained from Table 8.1 by interpolation, assuming the surface pressure to be about 1000 mb.)

7   What is the difference between cloud drops and rain drops? Discuss the ways in which rain drops can form in a cloud.

8   Why doesn't a falling large drop in a cloud collect every smaller drop in its path?

# NINE

## Convection. Cumulus Clouds, Thunderstorms, and Tornadoes

### 9.1 The development of convective clouds

In Chapter 1 a preliminary description of the development of convective clouds was presented. We are now in a position to discuss this phenomenon more completely. We know, for instance, that on a clear morning the heating of the ground by solar radiation leads to destabilization of the layer of air next to the ground, that is, to the development of a lapse rate $\gamma$ greater than the dry adiabatic lapse rate $\Gamma$. Due to unevenness in the amount of heating because of differences in the character of the surface, or due to differences in flow, disturbances occur that release the instability and produce upward and downward motions. These motions are organized in either *cells* or *rolls*, the nature and scale of which depend on the rate of heating and the prior state of flow of the air.

That vertical motions start before clouds develop in them is readily seen at times. If you are hiking on a dusty trail with the sun beating down on you, the discomfort is increased by the clouds of dust that are stirred up by those ahead of you on the trail and carried upward in rising currents of heated air. Similarly, the ploughing of fields or trucks traveling on dirt roads raise swirls of dust to considerable heights, and dust devils, small whirlwinds that are rendered visible by the dust rising at their center, occur over heated fields, particularly in the desert. All these provide evidence of upward motions caused by thermal instability.

**Figure 9.1** *Lightning discharges in a thunderstorm. [Courtesy of NOAA.]*

The nature of the organization of these dry thermals into cells and rolls was pointed out by Alfred Woodcock a number of years ago. He observed that sea gulls seek out the rising air currents to soar in. By remaining in the updrafts they can soar gracefully for long periods without moving a wing. Woodcock noticed that when the wind is light the gulls soar in circles, but when the wind exceeds a certain critical value (about 6 ms$^{-1}$) the pattern of their soaring changes to one in which they travel in lines along the direction of the wind. He interpreted this change in soaring pattern in terms of the type of pattern of convection—cells at low wind speeds and rolls oriented along the wind when the winds are sufficiently strong.

If the rising air is humid enough it will reach the condensation level and clouds will form in the shape of cells or helical rolls in accord with the structure of the convection that is present. Depending on the stability of the air above the layer that is heated from below, the clouds will be shallow or deep. If the lapse rate there is less than the saturation adiabatic ($\gamma < \Gamma_s$) only *cumulus humilis*, "humble" cumulus clouds of fair weather, will occur. If $\gamma > \Gamma_s$ through a deep layer, the cumulus will grow upward into cumulus congestus and ultimately into cumulonimbus, with lightning, thunder, heavy rain, and sometimes hail. In some instances the instability (with respect to saturated air) extends throughout the troposphere and the cumulonimbus tops reach the tropopause.

Figure 9.2 shows an instance of the growth of cumulus into cumulus congestus and cumulonimbus. The photographs were taken at intervals of about eight minutes from a distance of about 70 km. In photograph A cumulus clouds in the shape of a roll or a group of rolls are present. Photograph B shows some cumulus congestus towers extending upward at several places. In C, the towers have grown, but the motions are becoming more organized, with the two towers on the right dominating. In photograph D, the right-hand tower has become glaciated (the particles at the top have turned to ice), with characteristic smoothness having replaced the bulbous cauliflower configuration, and the other towers appear to be less energetic. In E and F the anvil shape is evident, and the entire cumulonimbus tower continues to rise. Throughout the development some small clouds remain at about the height of the original cumulus tops.

Investigations using airplanes that pass through the clouds at several levels at the same time have shown that a well-developed thunderstorm typically consists of several cells at different stages of their development. The stages of individual cells are shown in Figure 9.3. In the growing cumulus stage the motion throughout the cloud at all levels is upward. The cloud drops are still small enough to be carried upward with the air currents. As the cloud gets deeper and the top reaches levels where the temperature is low enough, ice crystals begin to form and the collision-coalescence growth of drops begins. In the mature stage the precipitation is under way, and with it a downdraft has developed within the cloud, induced by the drag of the falling drops. The updrafts and downdrafts reach speeds of 30 ms$^{-1}$ and are turbulent, so that planes flying through thunderstorms at this stage may be subjected to accelerations large enough to produce structural damage. Usually, pilots will try to avoid flying through cumulonimbus clouds both for the sake of the comfort of the passengers and from the standpoint of safety.

## 9.2 Influence of the environment. Entrainment

When heated air rises it cannot leave a vacuum, so air that has been surrounding the position it occupied near the ground must flow in to take its

A

B

C

**Figure 9.2** *Example of growth of cumulus into cumulonimbus. [Courtesy of University of Chicago Cloud Physics Laboratory.]*

D

E

F

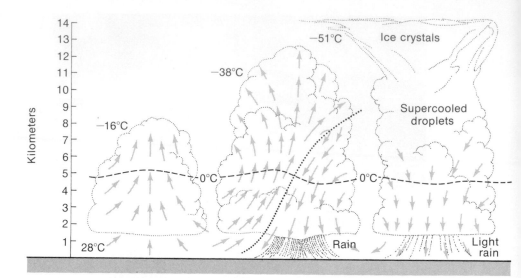

**Figure 9.3** *Stages in the development of a thunderstorm cell: left, cumulus or growing stage; center, mature stage; right, advanced or dissipating stage.*

place. Similarly, the air above it must move aside to make room for it, and there must be downflow to replace the air that has flowed in at low levels and to make room for the outflow aloft. Thus a complete circulation is established. The descending air current is usually much more widespread than the rising current, so that its speed is usually much lower. Nevertheless, the descent results in adiabatic heating, so that the temperature of the air around the convective cloud will be higher than the original environmental temperature, and the buoyancy force, which depends on how much warmer the cloud is than its surroundings, will be correspondingly reduced. The result is that the lapse rate must be considerably greater than the saturation adiabatic for numerous large convective clouds to develop. If the lapse rate is only slightly greater than $\Gamma_s$, the clouds will be widely scattered.

In addition to the effect of its sinking motion discussed in the preceding paragraph, the environmental air has an effect on the cloud growth because of the turbulent mixing that occurs at the sides and top of a growing cumulus. Unsaturated cooler air from outside is mixed into the cloud by this process, causing the drops either to evaporate or to grow more slowly and producing a decrease in the temperature of the cloud. The addition of environmental air into a cloud by mixing is called *entrainment*. As a consequence of entrainment the rate of decrease in temperature of the rising saturated air parcel is greater than the saturation adiabatic rate, and the buoyancy is correspondingly reduced. Thus both the sinking of the surroundings and the entrainment of environmental air decrease the difference in temperature between the cloud and the air around it, and thus reduce the rate at which the rising current is accelerated. The criterion for instability, which determines whether or not convection will begin, is unchanged by these influences, but they act as a partial negative feedback mechanism that makes it necessary that $\gamma$ be considerably larger than $\Gamma_s$ for widespread active thunderstorms to occur.

## 9.3  Processes that produce changes in stability

We have seen that the heating of the layers of air near the ground by the sun's radiation during the day makes these lowest layers of the atmosphere unstable (and, in turn, the cooling from below during the night renders them stable again). But this effect extends upward only 2 or 3 km at most, ordinarily. The stability of the layers at higher levels, which determines whether there will be deep convective clouds accompanied by showers and thunderstorms, depends on other factors. The factors are related to the general flow pattern of the air. The principal ones are *relative advection*, *horizontal convergence*, and release of *convective instability*.

Relative advection refers to the horizontal movement of air with different temperatures at different levels. Advection means simply the horizontal transport of air and its properties. Suppose that at lower levels the wind is bringing in warmer air, say, from the south, while at upper levels colder air than that which was present is being brought in, say, from the west. The effect of this warm advection below and cold advection aloft will be that as time goes on the temperature decreases more rapidly with height, that is, the lapse rate increases with time. If this process continues long enough, the air over a place could change from stable to unstable. The reverse type of relative advection, with cold air advection below and/or warm air advection aloft would tend to stabilize the air or increase its stability.

In horizontal convergence the air is moving with different speeds at the same level in such a fashion that air tends to accumulate in an air column as it moves along. This process is illustrated in Figure 9.4A, which shows the horizontal and vertical cross sections of an air column at two different times, together with the sounding curves for the two times on a thermodynamic diagram. The wind along $AC$ is assumed to be stronger than the wind along BD. After a time, the shape of the horizontal section of the column has changed to $A'B'C'D'$. To contain the same mass (at the same temperature and pressure) the volume must remain constant, and with the reduction in its horizontal area the vertical dimension of the air column must increase. Therefore $EF$ moves to $E'F'$. The lifting of the air above $E'F'$ leads to an outflow at higher levels so that the pressure at $E'F'$ stays at approximately the value it had at the same height before the convergence. In the thermodynamic diagram the sounding curve before the convergence is shown by line $AE$. During the convergence the pressure at the top of the column changes, and the air there goes through the

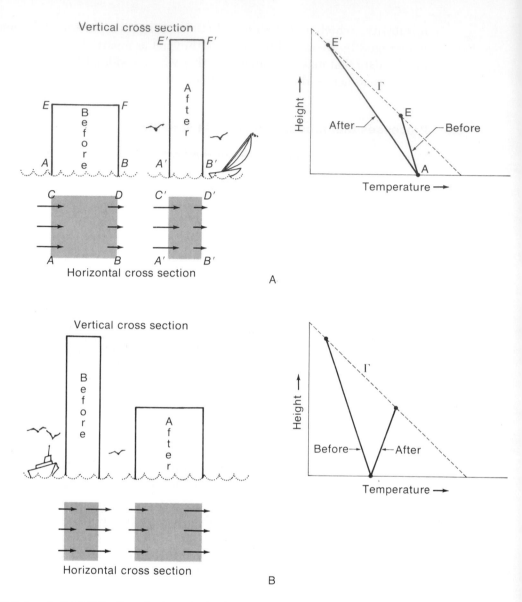

**Figure 9.4** *A. Destabilization of a layer by horizontal convergence and vertical stretching. B. Stabilization by horizontal divergence and subsidence.*

adiabatic process *EE'*. The sounding curve of the air column after horizontal convergence and vertical stretching is *AE'*. This process thus increases the lapse rate. However, it can never render an initially stable unsaturated air column unstable. If the lower part of the air column were initially saturated and the upper part unsaturated, however, the horizontal convergence and the accompanying upward motion of its upper portion could produce instability in the air column.

When an air column has a moisture distribution such that its vertical motion can produce instability, the column is said to be *convectively unstable*. For the convective instability to be released by horizontal con-

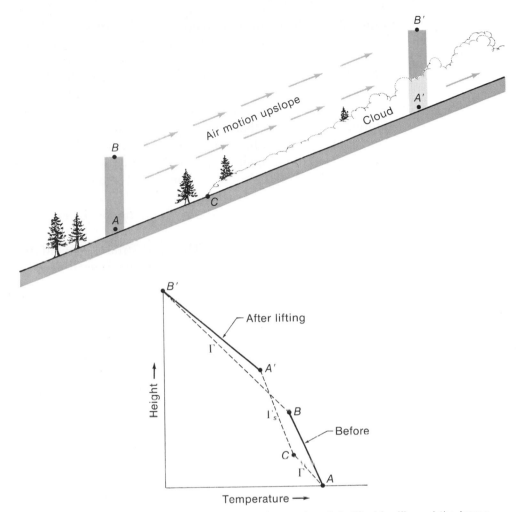

**Figure 9.5**  *Convective instability. The destabilization of a layer when it is lifted bodily and the lower part becomes saturated before the upper part does.*

vergence alone, it is necessary that the lower portion of the air column be saturated initially. However, if the entire column is lifted (with or without convergence) and the lower part is sufficiently humid to reach saturation before the upper part, instability could arise. This process is illustrated in Figure 9.5. In this diagram the air column $AB$ is pictured as moving up a mountain slope to $A'B'$. Its sounding is shown on the thermodynamic diagram by the curve $AB$. It is assumed that, in ascending, the lower portion of the column becomes saturated at $C$, but the upper portion remains unsaturated. The process curve for the bottom of the column is thus $ACA'$; for the top it is the dry adiabat $BB'$. The sounding curve after the ascent is $A'B'$. Since the lower portion is saturated and the lapse rate is greater than $\Gamma_s$, the column has become unstable.

Divergence can increase the stability of an air column in the same fashion that convergence can reduce its stability. In fact, inversions are frequently produced or reinforced by the divergence and subsidence associated with stationary or slow-moving anticyclones. The process of inversion formation by horizontal divergence is illustrated in Figure 9.4B.

In addition to the changes in stability produced by the influence of the earth's surface because of its warming by solar radiation during the day and cooling by radiative loss at night, air may pass over terrain that has a different temperature than the air and be rendered unstable or stable. For instance, if air moves in winter from snow-covered land onto an unfrozen lake or ocean, it will be heated rapidly from below and rendered unstable. This process leads to cumulus clouds and snow showers at the leeward shores of the Great Lakes in winter, with the amounts of snow frequently totaling several inches. An instance of the opposite effect is the air moving northward from the warm waters of the Gulf Stream to the very cold water off the coast of Newfoundland. The air is strongly cooled from below, producing an intense inversion. Dense fogs are caused by the condensation as the air is cooled.

Radiative exchange between layers of air aloft can also change the lapse rate. Ordinarily, these changes occur slowly and are small. An exceptional case is the radiational exchange at the top of a cloud. The liquid water in the cloud radiates nearly as a black body and loses more energy than the water vapor and $CO_2$ in the air layers above radiate back. This loss of heat may cool the upper part of the cloud sufficiently to restore or augment the instability and cause the cloud to grow. The occurrence of nocturnal thunderstorms is attributed partially to this process. The clouds develop during the afternoon, started by solar heating, but may not have quite enough energy to penetrate a stable layer aloft. The additional cooling at the cloud top may be sufficient to stimulate updrafts in the cloud that are strong enough to get through the layer that stopped its earlier growth.

To recapitulate, the stability of layers of air can be changed—

1  by heating or cooling from below due to solar heating or radiative cooling of the ground;
2  by heating or cooling from below due to passing over a warmer or colder surface;
3  by heating or cooling at upper levels by radiative exchange;
4  by relative advection;
5  by convergence or divergence;
6  by release of convective instability by convergence or lifting.

Frequently, more than one of these processes go on at the same time.

## 9.4 Relation of convective activity to flow patterns

The factors that contribute to destabilization through deep layers result in a relationship between larger scale flow patterns and showers and thunderstorms within air masses. Some of the air masses and circulations in which convective activity is favored are —

1　fresh polar air moving cyclonically over warmer land or water;
2　tropical maritime air moving poleward over land heated by solar radiation in spring and summer;
3　moist air subjected to convergence effects of sea breezes, valley-mountain winds, or monsoons.

The properties of polar air masses and the changes they undergo after leaving their regions of formation will be described in Chapter 10. For now it is sufficient to say that as they move equatorward they are heated from below and the portion of them next to the ground is rendered unstable. In general, equatorward motion is divergent, but cyclonic motion tends to be convergent. If the flow is sufficiently cyclonic the equatorward effect will be overcome and there will be enough convergence to destabilize the upper layers, allowing convective clouds to become thick enough to produce showers. Cyclonic flow of fresh polar air usually occurs immediately behind a cold front. Farther back, the flow straightens and becomes anticyclonic, leading to stability and clearing. Furthermore, the ground becomes cooled as the cold air moves over it, and after a while it no longer warms and destabilizes the air. Thus the showers in the cold air occur only a short distance behind the cold front.

The properties of tropical maritime air will also be discussed in Chapter 10. Without going into detail at present, therefore, we shall state only that, as the name implies, it is a warm, moist air mass. The moisture is greatest in the lowest layers; aloft the humidity is frequently quite low, so it is convectively unstable. When tropical maritime air comes over land in winter it is warmer than the ground and becomes stabilized; convection is suppressed and only fog or stratus cloud with drizzle is present. In spring and summer, however, the land heated by the high sun is warmer than the oceans from which the tropical air comes. The air is destabilized and showers and thunderstorms occur. In regions where the flow is poleward (i.e., south winds in the Northern Hemisphere), there is an additional destabilizing effect due to convergence, and the showers and thunderstorms are general. Where the flow is toward the equator or anticyclonically curved, divergence suppresses the convection and the showers are scattered or absent.

The tropical maritime air is usually associated with a large anticyclone. On the east side of the anticyclone, where the flow has a northerly component (Northern Hemisphere), there is little or no convective activity; to the west of its center, the southerly flow, with its convergence, produces widespread showers and thunderstorms.

Sometimes the thunderstorms in the tropical air are organized into a line of convective activity called a *squall line*. The line in this case is not

**Figure 9.6**  *Tornado at Enid, Oklahoma, June 5, 1966. [Photograph by Leo Ainsworth, Courtesy of NOAA.]*

simply the roll type of convective motion discussed at the beginning of this chapter. Even a single large thunderstorm may have strong winds associated with the outflow of cool air arising from the downdraft. Detailed analysis has shown that the intense convective system, including the downward motion produced by the precipitation, gives rise to a small high-pressure center that pushes air that has been cooled by the evaporation of precipitation falling through it outward ahead of the storm in the form of a "pseudofront." This outflow releases the convective instability of the air ahead of it, thus propagating the convection. When several thunderstorms occur side by side, the outflow forms a continuous line of strong gusty winds.

Squall lines may originate entirely within the tropical air mass, or they may be set off by the impetus of the cold fronts to the west. In the latter case, they frequently move faster than the fronts and are soon no longer directly associated with them. With the arrival of a squall line the wind usually shifts direction abruptly, as well as increasing in speed. The temperature drops, and heavy rain begins to fall, accompanied by lightning and thunder and sometimes by hail. Hailstones as large as 10 cm across have been observed, although usually they are about 1–2 cm or less in diameter. The winds accompanying severe thunderstorms may reach 30 m s$^{-1}$ or more and cause damage to structures and to trees. The most intense windstorms of all, the tornadoes, are associated with severe thunderstorms arising from squall lines.

Regions where convergence occurs frequently, as a result of winds produced by topographic effects, will have convective clouds and showers if other conditions are favorable. For instance, the Florida peninsula has sea breezes blowing into it from three sides, producing convergence that destabilizes the air and leads to daily showers in summer when moist air is present and the heating by the sun helps set off the convection. Consequently, the region of the United States in which thunderstorms are most frequent is Florida. A secondary maximum in the frequency of thunderstorms is found in the southern Rocky Mountains, where the valley-mountain winds produce upslope motion and convergence over the mountains. The thunderstorms over the mountains of New Mexico and Colorado occur mostly during the afternoon. Over the Great Plains to the east the flow toward the mountains causes divergence during the day, but the outflow from them at night causes convergence. This diurnal pattern of convergence and divergence, together with relative advection effects that are similarly influenced by the mountains and possibly the radiative exchange at the cloud tops, causes thunderstorms to occur more often at night than during the afternoon over the Great Plains. Monsoon circulations have similar effects on the occurrence of convective showers, with seasonal instead of diurnal periods.

## 9.5 Tornadoes

Tornadoes represent an extreme culmination of convection, in which the kinetic energy becomes concentrated into an intense vortex or whirlwind. They have the strongest winds occurring at the earth's surface, with speeds estimated to reach up to 120 m s$^{-1}$. They are always associated with severe thunderstorms, and usually with squall lines. They are quite small in extent, ranging from 100 m to about 1 km in diameter, but are locally the most destructive of all storms, with almost all structures and trees in their paths suffering complete devastation.

The characteristic funnel cloud of a tornado (Figure 9.6) appears to be drawn down from the cumulonimbus cloud of the parent thunderstorm, but actually the air rotating around the storm moves inward toward the center and then upward. The dust and debris lifted from the ground, including large objects such as cars, shows the strength of the upward current. The pressure at the center is much lower than at the periphery, with decreases in excess of 50 mb having been measured and values up to 100 mb estimated. The funnel cloud shows the position where the moist air flowing inward toward the center becomes saturated by adiabatic cooling.

The rapid pressure fall when the center of a tornado passes over structures produces an explosive effect which adds to the destruction due to the strong winds. Unless buildings have enough open doors and windows to permit rapid equalization of pressure, the force due to the excess pressure inside the building literally raises the roof and pushes out the walls, which the wind then carries away. Injuries and fatalities in tornadoes are mostly due to the victims' being crushed by the collapse of structures or struck by objects flying in the wind.

Usually the funnel cloud of the tornado extends downward and reaches the ground for only a few minutes, during which it moves a kilometer or two. However, the small number of tornadoes that cause deaths characteristically remain at the ground for distances ranging between 20 and 50 km, suggesting that the tornadoes that have longer lives and move farther are the more intense ones. A typical tornado moves northeastward with a speed of translation of the vortex of about 20 m s$^{-1}$, leaving a path of destruction about 400 m wide. (Note the distinction between the speed of the tornado, about 20 m s$^{-1}$, and the speed of the winds rotating around its center, up to 200 m s$^{-1}$.) On rare occasions tornadoes have been reported to remain practically stationary, and occasionally they have traveled with speeds up to 30 m s$^{-1}$. Their path is usually along or parallel to squall lines or cold fronts, but frequently they occur in other situations, for instance, in association with hurricanes. The rotation of tornadoes is usually cyclonic, although instances of anticyclonic rotation have also been reported. In appearance tornadoes frequently have a double structure, the true funnel cloud being surrounded by a separate cylinder of dust and debris raised from the ground.

While tornadoes have occurred in every state of the United States and in many other parts of the world, they are most frequent in the central and southeastern United States, with maximum frequency in a band extending from the Texas Panhandle northeastward to northwestern Missouri. In a typical year, between 500 and 1,000 tornadoes occur in the United States. The total number reported in the eleven-year period 1960–1970 was 7,428. Of these, 223 tornadoes had fatalities associated with them, producing 1,014 deaths.

In the late winter the most probable location for tornadoes is the southeastern states. As spring progresses they are more likely farther west and north; in summer they occur in the Great Plains and the northern states, but with the decrease in air-mass contrasts, they become less frequent. More than half of the tornadoes in the United States have occurred in

April, May, and June; of those that have occurred in Kansas and Oklahoma, three-quarters took place in these three months. However, there is no month of the year in which tornadoes have not occurred.

With respect to time of day, tornadoes, like thunderstorms, are most frequent in the afternoon. About one-third of those that have occurred in the United States have taken place between 3:00 P.M. and 6:00 P.M., local time, and 82 percent have taken place between noon and midnight.

The general conditions favorable to severe thunderstorms and tornadoes can be forecast fairly well twelve or more hours in advance. However, specifying the exact time and place of occurrence of individual storms, affecting, as they do, very small areas, defies present methods of prediction in advance of their development. A network of radar stations has been established that covers much of the area subject to frequent occurrence of tornadoes. When a tornado is either seen or detected by radar, prompt warnings are issued by radio and television to alert the people in the vicinity.

Over the ocean tornadoes are called *waterspouts*, because of the lifting of water at their base by the wind and updrafts.

## Questions, Problems, and Projects for Chapter 9

1 Discuss the influence of the environment of a convective current on it.

2 Why is it necessary that $\gamma$ be considerably larger than $\Gamma_s$ for deep convection and widespread active thunderstorms to take place?

3 Discuss the kinds of weather that are characteristic of the eastern and western portions of a subtropical anticyclone and the reasons for the difference.

4 Why are the maximum frequencies of thunderstorms in the United States in Florida and over the Rocky Mountains?

5 Describe two weather situations in each of which more than one of the processes that are listed at the end of Section 9.3 act to change the stability of the air. Explain how these processes act and what the effect of the changed stability is.

# TEN

## Air Masses

### 10.1   The nature and classification of air masses

If air remains for a long enough time over a portion of the earth's surface, its properties tend to become typical of that surface. If the character of the surface is approximately the same over a large area—for example, a broad expanse of warm ocean—the properties of a large body of air will become nearly uniform in the horizontal. Such a body of air, with properties (in particular, temperature and moisture content) nearly uniform over horizontal distances of the order of thousands of kilometers, is called an *air mass*. The areas where they form are called *source regions*.

For an air mass to form, the air must stagnate over large areas with uniform properties. In general, this means areas dominated by stationary or slow-moving anticyclones, since anticyclones usually have extensive areas of calms or light winds. For the formation of air masses light winds and uniform surface properties must be present together. The light winds insure that the air will stay over the source region long enough to come approximately to equilibrium with it. The principal characteristics of source regions that determine the air-mass properties are their temperature and whether they are land or sea. As indication of their temperature, source regions are classified by latitude belts—equatorial, tropical, polar, and arctic (or antarctic)—and the surface character is designated as continental or maritime. Based on these categories of the source regions, air masses are classified as follows:

| Arctic | A |
| Maritime Polar | mP |
| Maritime Tropical | mT |
| Continental Polar | cP |
| Continental Tropical | cT |
| Equatorial | E |

The abbreviations are sometimes used on weather maps to show the air mass over an area, particularly when the air masses have moved from their source regions. Arctic air is typically continental in its properties, particularly in winter, and equatorial air is typically maritime.

## 10.2  Properties of air masses

The properties of the air masses, being derived from their source regions, need almost no elucidation. The continental air masses are generally dry, at least in absolute moisture content, while the maritime air masses have high humidities. The equatorial and tropical air masses are warm, the arctic and polar (by which is really meant subpolar) are cold.

Arctic air masses are present principally in winter. They form in the polar anticyclone and are characterized by very low temperatures. Their absolute humidity is very low, because of the low temperature, but their relative humidity may be high. They are very stable near the ground, with strong inversions extending through the lowest kilometer or two.

Continental polar air masses in winter are similar in properties to arctic air, but not quite as cold. They form, in the Northern Hemisphere, when anticyclones stagnate over Alaska, Canada, Russia, and Siberia. In summer they are moderately cold, with more variable humidity and much less stability.

Maritime polar air forms over oceans at high latitudes. While air rarely stagnates in these regions, the source regions are sufficiently extensive so that the air can attain their characteristic properties even though it is moving fairly rapidly. In winter mP air is relatively mild, compared with cP or A; in summer it is cool; in both seasons it is moist and conditionally unstable.

Continental tropical air forms over subtropical land areas: North Africa, southwestern United States and Mexico, and the desert areas in Asia (particularly in summer). It is hot and dry. It is unstable, but clouds are rarely present in it because of the very low humidity.

Maritime tropical air forms over the oceans at low latitudes on the equatorial side of the subtropical anticyclones. It is warm, moist, and unstable in the lower layers, but at a short distance above the sea there is an inversion, above which the air is hot and dry because of subsidence. As the mT moves around the anticyclone and starts poleward over the western portions of the oceans, the moist and (conditionally) unstable layer deepens.

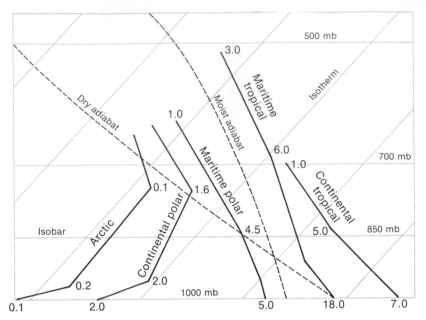

**Figure 10.1** *Schematic sounding curves for various air masses, plotted on Skew T-Log P Diagram. Numbers next to sounding curves are mixing ratios in parts per thousand.*

Equatorial air forms near the intertropical convergence zone. It is warm and moist near the ground and unstable to high levels except over the eastern parts of the oceans where the sea surface is relatively cold because of the upswelling of deep water.

In Figure 10.1 are schematic sounding curves representing the typical variation of temperature and humidity of the various air masses with height in their source regions.

When air masses move from their source regions they are modified by interacting with the surfaces over which they pass and by the various processes described in Section 9.3 that alter their stability. They also may interact with each other. Their interactions with each other are described in Chapter 11. Some of the transformations they undergo are discussed in the next section.

## 10.3  Transformation of air masses

When arctic and continental polar air masses move equatorward they are usually heated from below and thereby rendered unstable. Frequently, evaporation from warm water surfaces or from moist soil and vegetation increases their moisture content. Consequently, convective clouds develop in them. If incorporated in a cyclonic flow, there may be convergence that contributes to the development of instability through deeper layers, enabling the clouds to become large enough for showers to occur.

More often, however, southward flow (in the Northern Hemisphere) is anticyclonic and accompanied by divergence that gives rise to a stable layer that limits the height to which the clouds grow even though the air continues to be warmed from below.

When A or cP moves out over the Atlantic or Pacific Ocean the rate of heating is so great that the instability and moisture are quickly transported to considerable heights and the air is rapidly transformed to mP. Sometimes the air crossing the ocean is caught up in the circulation around the sub-tropical high and, passing over the warmer portion of the ocean, reaches land again as mT.

If mP moves from the Pacific over North America, it must ascend the mountain ranges of the west. Beginning with the coastal ranges, this ascent causes cloudiness and precipitation. Since the mP air usually has convective instability that is released by the ascent, the precipitation is heavy. On descending the mountains, the air is heated by adiabatic compression and reaches the plains to the east much warmer and drier than it had been. When mP replaces cP over the Great Plains in winter it brings mild temperatures, frequently with clear weather, after the bitter cold and snow showers of the cP. Sometimes the mild mP air arrives suddenly, with strong winds, and produces increases of temperature of as much as 20°C in 15 minutes. As discussed in Chapter 5, the warm, dry wind descending the east slope of the Rockies is called the *chinook*. A similar warm wind descending the Alps is called the *foehn*. In Southern California the warm, dry wind descending from the elevated terrain inland is called the *Santa Ana*, but the Santa Ana consists of air that has crossed the Coast Ranges and the Sierra Nevada some days earlier and participated in subsidence over the Great Basin before turning southwestward, descending from the plateau and flowing down through the passes in the mountains to reach the coastal plains.

The transformation of moist air as it ascends over a mountain range and is heated dry adiabatically as it descends is illustrated in Figure 10.2. The precipitation occurs principally on the windward side. Sometimes on the lee side a standing wave develops in the flow, in which clouds form as the air ascends and dissipate as it descends. These clouds, like those over the windward side of the mountains, remain in place as the air passes through them. Because they frequently are lens shaped, they are called *lenticular clouds* (Figure 10.3). Occasionally, several lenticular clouds occur above each other, like a stack of pancakes, or other odd and spectacular arrays of them are present.

Tropical maritime air moving poleward over the ocean or over continents in winter tends to be rendered more stable, and fog or stratus clouds develop in it. However, when mT air moves over land in summer (and throughout the year at low latitudes), it may be rendered unstable, since the land surfaces are warmer than the oceans. Cumulus clouds with

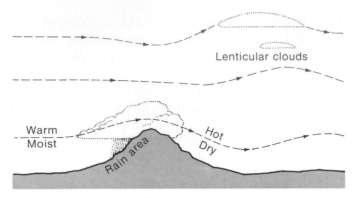

**Figure 10.2**   *Transformation of mP air flowing over mountain range.*

showers and thunderstorms occur in mT moving poleward over land in summer.

The structure of air masses is also changed by the convergence or divergence associated with large-scale flow patterns. The effect of convergence and divergence on the stability of polar continental air masses has already been mentioned. Divergence is associated with equatorward motion and anticyclonic flow, which therefore tend to stabilize the air and produce inversions, while poleward and cyclonic flow are convergent and thus tend to reduce the stability. The effects of these motions are felt principally at higher levels, since the amount of vertical motion (vertical stretching with convergence and shrinking with divergence) is proportional to the thickness of the column in which the convergence or divergence occurs.

An outstanding example of the effect of divergence on air-mass structure is the flow in the eastern portion of the semipermanent HIGHS over subtropical oceans. Here equatorward motion combines with anticyclonic flow to produce an extensive area of persistent divergence. The result is that the air descends from the middle troposphere and is heated by compression, producing the trade-wind inversion discussed near the end of Chapter 4. The air from high levels cannot descend all the way to the sea surface, and, in fact, the air near the sea gains heat and moisture as it moves equatorward over warmer water, so that a shallow unstable "marine" layer is formed, topped by the intense inversion. The depth of the marine layer depends on the relative intensity of the heating from below and the subsidence of air from above. Near the continents to the east, where it is shallow, the marine layer contains stratus clouds or fog. Farther west, where it is deeper, characteristic "trade-wind cumulus" are present. Figure 10.4 illustrates the process of modification of mP air to form the trade-wind inversion.

As a final example of the modification of air masses, the process by which mP air is transformed to cP or A air when it moves from the open oceans to

**Figure 10.3** *Lenticular clouds formed in wave in lee of mountains.* [*Courtesy of NOAA.*]

a snow-covered continent or the ice-covered Arctic Ocean will be discussed. When the relatively warm maritime air begins to stagnate over the frigid land the lowest levels are cooled by conduction and radiational exchange, establishing a shallow but intense inversion (Figure 10.5). The increased stability keeps the cooling from being "carried upward" by convective or turbulent transport of heat from above. Further adjustment of the temperature occurs only by radiative exchange.

Even though snow is highly reflective in the visible, as shown by its intense white color, at the infrared wave lengths at which most of the energy is radiated at its temperature it radiates nearly as a black body. Since air emits only in the wave lengths at which water vapor and $CO_2$ absorb radiation, the return radiation from the layers of air up to the top of the inversion and above it, even though they are at a considerably higher temperature, will be much less than that emitted by the snow. Thus the snow will continue cooling until it reaches a temperature $T_s$ at which the emission from the snow is equal to the return radiation. If the air temperature were to remain constant, the snow surface would approach the equilibrium temperature $T_s$ and stay at that temperature. However, when the snow reached that equilibrium, the air temperature would not be in equilibrium.

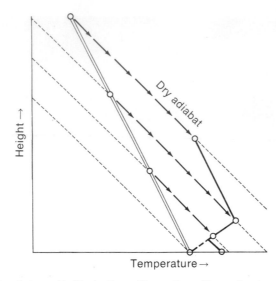

**Figure 10.4** *Schematic illustration of formation of inversion by horizontal divergence and vertical shrinking of an air column: double line shows the initial sounding; heavy line, the sounding after divergence, with the dashed portion of the inversion replaced by adiabatic layer because of heating at sea surface.*

Consider the layer of air in the vicinity of the inversion top in sounding curve 1 of Figure 10.5. It would be emitting radiation in those wave lengths at which it radiates at a temperature of about 270°K; the radiation reaching it from above and below at the same wave lengths, which it could absorb, comes from air or snow at lower temperatures. Thus it would be losing more energy than it gains from radiative exchange, and it would be cooled. Once it cooled, the snow surface would no longer be in equilibrium and would start cooling some more. The cooling would affect progressively thicker layers of air, as shown by curves 2 and 3 in the figure. In this fashion the mP air is transformed to A or cP air through depths of several kilometers by radiative cooling.

We have not considered short-wave radiation from the sun in this discussion. Since the albedo (reflectivity) of snow for solar radiation is more than 80 percent, and the intensity of solar radiation at these latitudes is small except at the time of the summer solstice, the role of solar radiation is negligible.

## 10.4 Air-mass weather

The weather within an air mass depends on the air-mass properties, particularly its humidity and stability, and on the flow characteristics. In the previous section we have seen several examples of this: cloudiness and

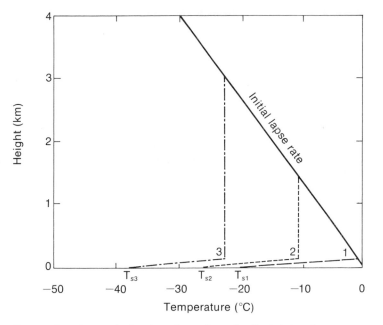

**Figure 10.5** *Successive stages during transformation of mP air into cP air over snow in winter.*

precipitation resulting from humidification and destabilization as air masses moved from their source regions.

In general, maritime air masses tend to have cloudiness and showers, and continental air masses tend to be clear. When mT air moves poleward over colder water or onto colder continents in winter it is stabilized and fog or stratus clouds occur in it, and when cP air moves equatorward it is moistened and destabilized, and convective clouds or showers may occur, particularly if the flow becomes cyclonic.

Air masses frequently coincide in extent with stationary or moving anticyclones. In the eastern portion of the anticyclones the flow is equatorward and is usually divergent. This combination ordinarily results in stabilization of the air masses and clear weather. In the western portion of the anticyclones the flow is poleward and may be convergent. On the western side of anticyclones we would expect the air masses to become less stable and convective cloudiness and showers to occur. Thus with a single air mass there may be enough variation of horizontal properties for clear weather to be present in one part and showers to occur in another, particularly when the air mass has moved from its source region.

While much of the time the weather at a place is determined by the properties of the air mass that overlies it, the most severe weather usually is associated not with a single air mass but with the interaction of two air masses in the vicinity of the boundary between them, that is, at *fronts*. In the next chapter we shall examine some aspects of this interaction.

## Questions, Problems, and Projects for Chapter 10

1 a List the source regions from which you would expect air masses to reach your locality.

b For each such air mass, and separately for summer and for winter, describe (i) its properties at the source, (ii) the modifications it would experience in traveling to your locality, and (iii) the kind of weather your locality would have when it is present.

2 a In what parts of the world would you expect continental polar air to form in winter?

b Discuss the changes you would expect continental polar air to experience in moving over the ocean.

3 Discuss the kind of weather you would expect on the western side of the coastal mountains as maritime polar air crosses them in winter, and the kind of weather you would expect east of the Rocky Mountains when maritime polar air reaches there.

# ELEVEN

## Fronts
## and
## Cyclones

### 11.1 Introduction

The association of decreasing pressure with the approach of bad weather and rising pressure with clearing has long been recognized. In fact, following the custom started in the eighteenth century, many barometers still have legends on them with the word "Fair" at high values of pressure and "Storm" at low values. In general, this association is valid: the approach of centers of high pressure (anticyclones) brings fair weather, while cloudiness and precipitation come with cyclones.

In the early years of weather forecasting this relationship was the dominant consideration, with pressure changes being concentrated upon rather than the motion and interaction of air masses. However, even then some meteorologists attempted to understand the physical causes of the weather by constructing cyclone models that contained some of the features of modern theories of storm development. For instance, Heinrich W. Dove, in 1837, concluded that temperate-latitude weather systems were associated with the interaction of warm and cold air currents, and Robert Fitz-Roy, in 1863, published a diagram showing cyclones forming at the zone of interaction between warm, moist air coming from subtropical latitudes and cold, dry air coming from polar regions. These early studies were largely ignored by practicing forecasters until the formulation of the Polar

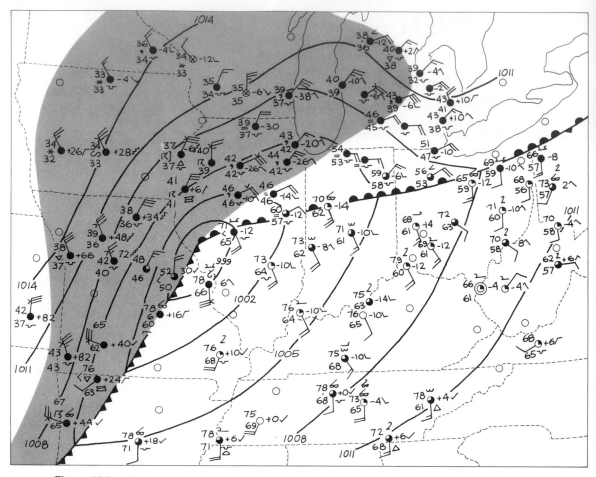

**Figure 11.1** *Map of surface weather over middle western United States, showing example of frontal wave. Shading indicates area of current precipitation.*

Front Theory of Cyclones by J. Bjerknes in 1918. This theory provided a systematic basis for understanding and forecasting the development of cyclonic storms in middle and high latitudes. The theory has been extended and modified through the years, but it remains one of the important guides to understanding weather patterns and the dynamic processes associated with them.

## 11.2 Fronts

The starting point of the polar front theory is that the boundary between two air masses approximates an abrupt discontinuity, rather than a wide zone of gradual transition. This boundary, of which the intersection with the ground shows up as a line on the weather map, is a surface, not vertical but sloping in such a way that the warm air extends above the cold air. On this frontal surface waves form that increase in amplitude and become cyclones, with low pressure at the center and winds circulating counterclockwise (Northern Hemisphere) around them.

In actuality, rather than a mathematical surface (of zero thickness), fronts are narrow zones in which the rate of change of air-mass properties with distance is large. Typically, the frontal zones of rapid transition from one air mass to the other are of the order of 100 km in width. Nevertheless, these zones permit wave behavior similar to that occurring at the interface between two fluids with differing densities and velocities. The nature of waves in "baroclinic" zones (layers in which the temperature structure varies rapidly in the horizontal) will be discussed briefly in the next section, along with the more readily visualized waves at a mathematical interface.

Figure 11.1 shows a portion of a weather map on which a typical wave cyclone appears. The fronts are shown by dark lines with solid half-circles or triangular points. The contrast in temperature across the front shows distinctly. For instance, in Illinois, just east of the LOW the temperature (in °F) is in the 70's south of the front and in the 40's north of it. Similarly, in Missouri, southwest of the LOW, the temperature east of the front is 78°F; west of it, it drops rapidly to about 38°F.

Temperature is the principal property used to identify air masses and to locate the fronts between them. Other properties that change at fronts and help the weather-map analyst locate them are the following:

1   The humidity, represented by the dew-point temperature.
2   The pressure gradient. As a consequence of the sense in which the pressure gradient normally changes across a front, isobars crossing the front bend in such a way that the vertices of the angles they form point toward high pressure. Frequently, the front lies in a trough of low pressure.
3   The wind direction and/or speed. The change is such that *cyclonic shear* is present at the front. By cyclonic shear is meant horizontal variation of the wind in such a sense that a small volume of air at the front would be rotated cyclonically by the wind.
4   Cloudiness and precipitation. The nature of the cloud and precipitation patterns associated with fronts and frontal wave cyclones will be discussed later in this chapter.

In general, fronts can occur between any two air masses that form side by side or are brought together by their motions. However, the air circulations in low latitudes are such that fronts do not ordinarily develop between equatorial and tropical air. The principal front from the geographical standpoint is the *polar front,* which is the boundary between polar and tropical air. It is not continuous all the way around the earth, but is interrupted by regions where the transition between tropical and polar air is gradual. The front between arctic and polar air masses is called the *arctic*

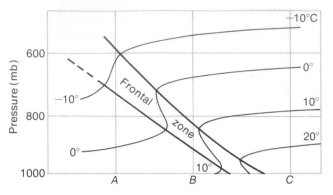

**Figure 11.2** *Schematic vertical cross section showing temperature distribution across a frontal zone.*

*front.* It, too, is not continuous around the earth. The fronts move with the air masses, and they undergo changes in sharpness and intensity as the air masses are transformed and the flow fields change. The average positions of the polar and arctic fronts undergo seasonal variations. In North America the average position of the polar front in winter is slightly south of Florida (extending northeastward toward the British Isles), while in summer it is north of the Great Lakes. Since cyclonic activity is principally associated with the polar front, there is a similar seasonal shift in the zones of maximum cyclone formation.

Fronts frequently occur between mP and cP air, or between an mP air mass that has undergone considerable modification and a fresh mP air mass.

When a front is moving it is called a *cold front* if cold air is replacing warm air at the ground (or sea surface), and a *warm front* if warm air is replacing cold air. If the air masses are moving parallel to the front so that the front does not move, it is called a *stationary front.* On manuscript weather maps cold fronts are represented by blue lines, warm fronts by red lines, and stationary fronts by lines composed of alternate red and blue segments. On printed maps cold fronts are represented by lines from which solid triangular points extend into the warm air; warm fronts are represented by lines with solid semicircles extending into the cold air; and stationary fronts by lines with alternate triangles on the warm air side and semicircles on the cold air side. In Figure 11.1 the front extending eastward from the LOW is represented as a warm front and the one extending southwestward from it is a cold front.

Figure 11.2 shows a schematic vertical cross section through a front with isotherms drawn to represent the temperature distribution. The front is shown as a zone of rapid change in temperature. The temperature decreases about 10°C through the frontal zone, and continues to decrease in the cold air. The horizontal temperature contrast is somewhat smaller at higher levels than near the ground, and at high levels the temperature decreases in the vertical through the front, whereas at low levels the fron-

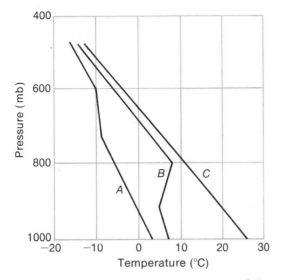

**Figure 11.3**  *Temperature soundings at points* A, B, *and* C *in Figure 11.2.*

tal zone is an inversion. This difference is shown in Figure 11.3, in which sounding curves corresponding to positions *A*, *B*, and *C* in Figure 11.2 are shown. At position *B*, where the front is low, there is a strong inversion, but at position *A* the frontal zone is represented only by a slightly more stable lapse rate than below or above.

The steepness of the front is greatly exaggerated in Figure 11.2, in which the vertical scale is much larger than the horizontal. In actuality, fronts are inclined upward toward the cold air side with slopes in the range 1:50 to 1:300. By this is meant that if one went 50 to 300 km into the cold air from the surface position of the front one would find the frontal surface at a height of one kilometer. Because friction slows the air near the ground more than the air higher up, cold fronts tend to be steeper and warm fronts less steep at low levels than aloft. The average slope of cold fronts is about 1:100; of warm fronts about 1:200. Fronts thus make very small angles with the horizontal. Nevertheless, this small angle is enough to provide the lift to the warm air flowing over the cold air to produce clouds and precipitation. Flow of the warm air up over the cold air develops typically in connection with frontal waves. In the next section we shall examine the nature of the frontal wave cyclone.

## 11.3  Frontal waves. The wave cyclone

At the surface between two fluids of different densities moving with different velocities waves are likely to be present. The most familiar example is the air-water interface at the top of lakes or oceans. When the wind blows over a water surface, waves develop. These waves have various lengths and amplitudes, and the speed of their movement depends on these factors (and the water depth if it is shallow) and is different from that of the wind or the water.

By analogy, it is natural to expect waves to occur at the surface between two air masses of different densities. As visible evidence of one instance of such waves, billow clouds, the wave-shaped altocumulus or cirrocumulus clouds, are seen frequently. It has been shown that these clouds are due to wave motions at an inversion between warmer air above moving relative to the cooler, moist air below. The clouds form with the upward motion of the wave, and the clear spaces are in the places where the air has descended enough to be heated adiabatically sufficiently to cause the cloud to evaporate. Billow-cloud waves have wavelengths of the order of 1 km. The waves that are of significance in the formation of cyclones have lengths of the order of 1000 km. If waves of this length are unstable, they develop into cyclonic storms.

The concept of stability of a wave is similar to the stability of a parcel in an air column. If the frontal surface is perturbed (say, by displacing part of it slightly toward the cold air), the perturbation will tend to be propagated along the front as a wave. As it moves it may also tend to be damped out, or else to grow in amplitude. If it tends to be damped it is stable, if it tends to grow it is unstable, and if it moves along the front without change in amplitude it is neutral. Stable or neutral waves are of little interest, for their amplitudes ordinarily are too small for them to be noticed on the weather map. If waves of considerable length are unstable they develop into cyclonic storms, with significant weather phenomena associated with them, and contribute to the exchange of energy and momentum in the general circulation.

The existence of frontal waves was discovered in 1918 by J. Bjerknes, by careful examination of surface weather maps. Figure 11.4 shows the detailed model of the wave cyclone and the accompanying weather pattern that was published soon afterward by J. Bjerknes and H. Solberg. In the plan view, or weather map, the wave-shaped front is shown by the dashed line, with a *warm sector* protruding into the cold air that sweeps around it. There is a broad band of cloudiness and precipitation ahead of the warm front, caused by the warm air moving faster than the cold air and climbing up the frontal surface (compare Figure 1.5). At the cold front the precipitation is confined to a narrow band where the warm air is being pushed upward by the cold air. Farther back, the model shows the warm air above the cold frontal surface moving faster than the front, and thus descending and producing rapid clearing.

The similarity between the idealized model and the actual situation shown in Figure 11.1 is readily seen, as well as some differences between them. Differences are to be expected, since the form of the wave and the extent and character of the clouds and precipitation depend on the particu-

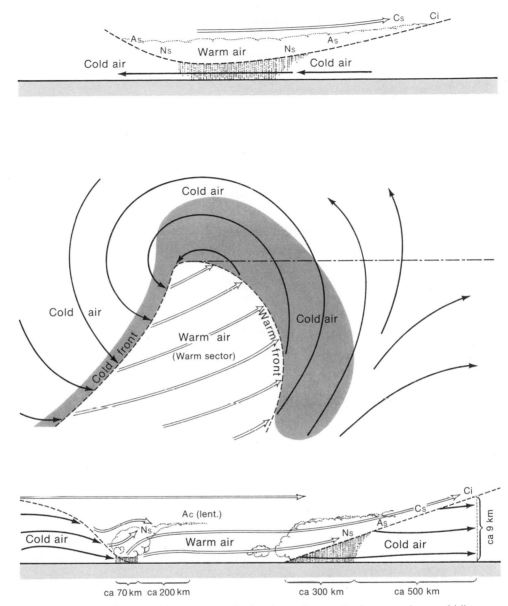

**Figure 11.4**    *Idealized wave cyclone model: upper—vertical cross section north of wave cyclone; middle—representation of frontal wave and streamlines on surface weather map; lower—vertical cross section through warm sector. Shading shows area of current precipitation. [From J. Bjerknes and H. Solberg,* Geofysiske Publikasjoner *(1922).]*

lar flow pattern and air-mass properties that precede the formation of the wave. If, as was the case in the example shown in Figure 11.1, the warm air ahead of the cold front is convectively unstable, showers and thunderstorms occur at the cold front. This situation frequently occurs in the central and eastern United States. At the warm front the upward motion is usually more gradual and the warm air is stable, so that the precipitation takes the form of light, continuous rain or snow over a wide area ahead of the front. Occasionally, the warm air ascending the warm front is convectively unstable and gives rise to thunderstorms there, too, and conversely sometimes the precipitation at cold fronts takes the form of continuous rain or snow in a wide band behind the front.

Consider the weather changes that would be experienced during the passage of a wave cyclone. Imagine yourself at the location represented by the middle of the extreme right side of Figure 11.4. The wave, with its accompanying weather, is moving from left to right. Initially, the wind would be from the south or southeast. The approach of the storm would show up first as a thickening cirrostratus cloud sheet approaching from the west. The pressure would fall—slowly at first, but steadily—and the clouds would become lower and thicker changing to altostratus. The cloud sequence as the warm front approached would be that described in Chapter 1. As the warm front moved to within perhaps 300 km, rain would begin to fall, at first light and intermittent, from clouds with bases near 3 km. Then the rain would become heavier and continuous, with the cloud bases lowering to 500 m or less. Fog would be likely as some of the rain evaporated into the shallow layer of cold air beneath the front.

With the passage of the warm front, the wind would shift to southwesterly, the temperature would rise, and the fog, precipitation, and clouds would disappear. This whole sequence would take place, on the average, in about 24 hours.

As the cold front approached, the pressure would begin to fall more rapidly. Some middle or high clouds might arrive as precursors of the line of cumulus or cumulonimbus that was approaching from the west. The arrival of the cold front would bring an abrupt shift of the wind to the northwest, a rapid decrease in the temperature, rising pressure, and showers or thunderstorms of greater intensity but shorter duration than the precipitation at the warm front.

If the pattern always conformed to the ideal model, it would be much easier than it is to predict the weather. As has been stated, differences occur because of differences in the conditions when the frontal wave forms. Topographic influences also cause differences in the characteristics of the waves. A large factor affecting the predictability of the weather is the fact that wave cyclones are not static entities that move horizontally without change, but dynamic organisms that undergo a "life cycle."

## 11.4  The life cycle of a cyclone. The occlusion process

Typically, the initial stage of a wave cyclone is a slight deformation of the polar front, frequently induced by a topographic feature. For this reason certain geographical regions are preferred areas for the formation of cyclonic waves, among them the areas east of the Rockies, in the vicinity of the Ozarks, and off the east coast of the Carolinas. If the wave is unstable

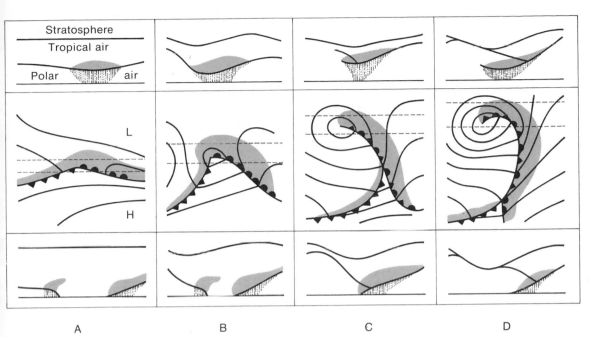

**Figure 11.5** *Stages in the life cycle of a wave cyclone: middle section repesents surface weather maps at various times; upper and lower sections, vertical cross sections at positions of dashed lines in middle section. A. Growing wave. B. Mature wave. C. Partially occluded wave. D. Wave near completion of occlusion process. [After C. L. Godske, T. Bergeron, J. Bjerknes, and R. C. Bundgaard,* Dynamic Meteorology and Weather Forecasting *(Boston: American Meteorological Society, 1957).]*

the frontal deformation becomes larger, with a bulge of warm air intruding into the cold air mass.

As the amplitude of the perturbation increases, more and more cold air is replaced by warm air, contributing to a decrease in pressure and the development of cyclonic flow around the region of low pressure at the tip of the warm sector. Within 24 hours of the initial disturbance of the front, a well-defined wave cyclone will have developed, with warm front, cold front, and pattern of cloud and precipitation corresponding to the ideal model in Figure 11.4. In an unstable wave the cold air behind the cold front moves faster than the air receding ahead of the warm front. The air in the warm sector is constantly being squeezed upward thereby. Eventually the cold front catches up with the warm front; at this stage the warm air is completely lifted from the surface and the cyclone is said to be *occluded*. The front that is formed by the merger of the cold front and the warm front is called an *occluded front*. The process of the cold air closing off the warm air from the ground is called *occlusion*.

Various stages of the life cycle of a wave cyclone and the occlusion process are shown in Figure 11.5. Diagram *A* represents the newly formed wave, with the pressure at the crest reduced just a little, and the precipitation bands quite narrow. In diagram *B* the wave is fully developed, with the characteristics described in the previous section. At this stage the air

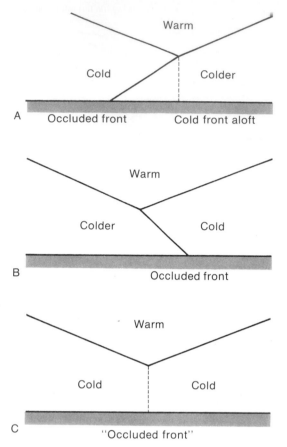

A  Occluded front       Cold front aloft

B            Occluded front

C        "Occluded front"

**Figure 11.6** *Representation of the types of occluded fronts: (A) warm-front type; (B) cold-front type; (C) occluded front with no temperature contrast at earth's surface.*

throughout the troposphere is participating in the wave motion, so that the tropopause is deformed, as shown in the vertical cross section at the top of the diagram. Diagram C shows the system after the wave has become partially occluded. The lifting of the warm-sector air lowers the center of gravity of the system, thereby converting potential energy into kinetic energy of the increased winds around the deeper low-pressure center. At a later stage (diagram D), the cold front has overtaken the warm front out to a larger distance from the center, practically eliminating the warm sector at the ground. The pressure at the center is still lower, with more closed isobars, forming a large vortex completely surrounded by cold air. After this, no further intensification can take place because the available supply of potential energy has been exhausted, and the cyclone gradually weakens and fills because of the action of friction.

In diagram C the vertical cross section through the occluded front (lower part of the diagram) has been drawn on the assumption that the cold air behind the cold front is warmer than that ahead of the warm front and

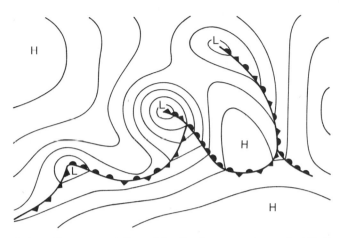

**Figure 11.7** *Schematic representation of family of wave cyclones at various stages on front.*

climbs up over it, leaving the front at the ground in the form of a warm front. In this situation the front is called a *warm-front-type occluded front*. It may occur when the air behind the cold front comes over the ocean while the air ahead of the warm front has had a trajectory completely over land, or when the cold air immediately behind the front has subsided from higher levels. The opposite situation frequently occurs, in which the air behind the cold front, coming more directly from the north (Northern Hemisphere), is colder than the air ahead of the warm front, which has spent a longer time since leaving its source region. In this case, the air behind the cold front pushes under the air ahead of the warm front, and the front is called a *cold-front-type occluded front*. Figure 11.6 shows vertical sections through the two types of occluded fronts, plus a vertical section through an occluded front with the same temperature on both sides. It should be noted that the precipitation comes principally from the warm air that is being lifted in a V-shaped trough in all three types, so that there is no difference in precipitation except with respect to the position of the precipitation in relation to the surface front.

In the warm-front-type occlusion the trough of warm air aloft is ahead of the front at the ground, at the position where the original cold front has climbed up the warm frontal surface. Since this position may be considerably ahead of the surface front and has significant weather associated with it, it is often shown on the surface weather map as a *cold front aloft*, represented by a dashed blue line or a broken printed cold front symbol. Occluded fronts are represented by purple lines on manuscript maps and lines with alternate semicircles and triangular points on the side towards which the front is moving on printed maps.

Frequently, a succession of waves forms on a front, so that a "family" of cyclones in various stages is present along the front. Figure 11.7 is a

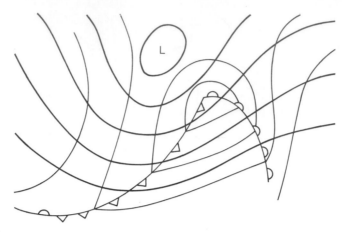

**Figure 11.8** *Schematic representation of relation between sea-level and upper-level flow in a wave cyclone. Heavy lines represent 500-mb contours; light lines are sea-level isobars and open front symbols represent the frontal wave there.*

schematic illustration of such a wave-cyclone family. Families of this type occur frequently, particularly over the Atlantic and Pacific Oceans. The wave cyclones reaching the American west coast from the Pacific are usually completely occluded members of such families. Many of the cyclones remain over the ocean, passing northeastward along the front as they occlude and merging with the semipermanent Aleutian or Icelandic lows.

## 11.5 Upper-level flow in relation to wave cyclones. Waves in the westerlies. The jet stream

When the models of the wave cyclone were first developed, observations of the temperature structure and flow patterns aloft were practically non-existent. Since then upper-air observations have clarified the relationship between the frontal waves and the flow at upper levels. At the same time, the observations have made it clear that other processes frequently lead to cyclogenesis (formation of cyclones) without the prior existence of fronts. Once the cyclone-forming process is under way, it may give rise to the formation of fronts within the cyclone. Thus there are two general ways in which cyclones may form: frontal waves at low levels that induce wave-shaped flow patterns aloft, and flow patterns aloft that produce low-pressure centers at lower levels and subsequently may draw air masses together, producing fronts in them.

In accord with the discussion in Section 6.6, the horizontal variation of the temperature in a frontal wave cyclone leads to variation with height of the pressure and wind fields. At upper levels the low-pressure center tilts toward the cold air, and since the isotherms on the cold-air side of the front are wave-shaped, some of the closed isobars tend to be replaced by wave-shaped isobars (or wave-shaped isobaric contours on the surfaces of constant pressure). Figure 11.8 shows schematically the upper-level pressure field (say, 500 mb contours) corresponding to the surface isobars in a wave cyclone. The low center at the upper level is displaced north-westward from the surface position, toward the cold-air side of the front, with only one closed isobaric contour around it. The rest of the flow, as shown by the contours, forms a wave, with the ridge ahead of the sea-level position of the warm front and the trough on the cold-air side of the cold front.

Maps showing the observed conditions at sea level and 500 mb on one particular day are shown in Figure 11.9. The general features just discussed are present in relation to each of the cyclonic systems: the low centers and troughs displaced to the cold-air side, and wave-shaped contours aloft replacing the closed sea-level isobars. The consequence is that the flow aloft (regarded as approximated by the geostrophic wind corresponding to the 500-mb contours) forms a wavy band across the entire map. In this instance there are three upper waves represented on the portion of the Northern Hemisphere covered by the maps, with troughs, each associated with an occluding wave cyclone, forty to fifty degrees of longitude apart. There would be seven or eight such waves going all the way around the earth.

Waves of this sort are characteristic of the westerly winds that prevail at upper levels in the troposphere. They vary in length. The shorter ones, associated with frontal waves at sea level, move rapidly, about ten to fifteen degrees of longitude per day. Longer waves, such that there would be four to six around the earth, move more slowly and are associated with the larger circulation features, such as the Aleutian and Icelandic lows and the subtropical highs. The flow in the long waves (known as *Rossby waves*, after C.-G. Rossby, a Swedish-American meteorologist who studied the dynamics of the upper waves) tends to "steer" the shorter waves, which move southward when behind the trough of a long wave and northward when ahead of it. In addition to changing direction, the troughs of the short waves intensify as they overtake the trough of a long wave and weaken as they catch up with the ridge.

Since the temperature gradient is large in the frontal zone the thermal wind in its vicinity is large, leading to a narrow band of very strong winds at high levels. This region of strong winds has been likened to a jet, for instance, the water spurting from a garden hose, and was named the *jet stream*. While its existence had been recognized earlier, it did not receive general attention until World War II, when bombing planes on their way to Japan encountered headwinds so strong that they could hardly make any progress relative to the ground. Winds as strong as 135 m s$^{-1}$ (300 mi/hr) have been observed in the jet maximum. The average speed in the center of the jet stream ranges from 35 to 55 m s$^{-1}$ at various longitudes in winter,

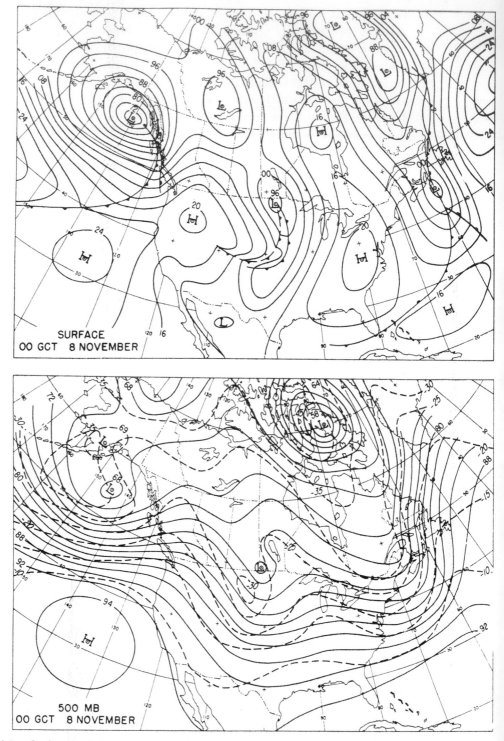

**Figure 11.9**  *Surface and 500-mb charts at the same observation time. Dashed lines on the 500-mb chart are isotherms, labeled in °C.*

and is somewhat lower in summer. The maximum speeds in the jet occur just below the tropopause.

Because other regions of concentrated high wind speeds have been found, the one associated with the polar front is called the *polar-front jet stream*. It has been likened to a meandering river winding its way from west to east at high speeds around the earth. Like the polar front, it is not continuous all the way around the earth, and sometimes it splits into two streams, which subsequently reunite. Being characterized by a large concentration of momentum and kinetic energy, it plays an important role in the dynamic processes of the upper troposphere.

## 11.6    Cyclogenesis

The mechanism by which cyclones are formed has been studied from several different approaches. In early theories, the development of low-pressure centers was thought to be due to heating and a convective circulation. These theories were disproven, except for tropical cyclones, when measurements of upper-air temperatures became available, which showed that cyclones in temperate latitudes had low temperatures aloft, and not the high temperatures required from hydrostatic considerations in order that low pressure may be present. The instability of frontal waves, as discussed in Section 11.4, provides one way in which cyclones form when the potential energy of position of the cold and warm air masses is transformed into kinetic energy of cyclonic circulation. Potential energy is likewise available whenever there is a variation of temperature in the horizontal, or more accurately, whenever the density varies in surfaces of constant pressure. Regions in which such a variation is present are said to be *baroclinic*, whereas regions in which the density does not vary along the surfaces of constant pressure are *barotropic*. Waves occurring in a baroclinic region may release potential energy and set up cyclonic circulations even when the baroclinicity is not concentrated in a narrow frontal zone. These waves may be set off in the westerlies of the upper troposphere, and the cyclonic circulation may work its way down to the ground. Sometimes when this occurs warm and cold air masses may be brought together by the induced circulation, that is, the baroclinicity becomes concentrated and fronts are formed. After the fronts have formed, the further development may resemble the life cycle of a frontal wave.

A large part of the problem of weather forecasting has been the determination of when and where cyclones will form and deepen, as well as in what direction and how fast they will move. Previous to the use of electronic digital computers, the forecaster depended a great deal on empirical rules that guided the early detection of incipient frontal waves and the estimation of their rate of development and their movement. By applying numerical methods of solving the equations of atmospheric motion, using

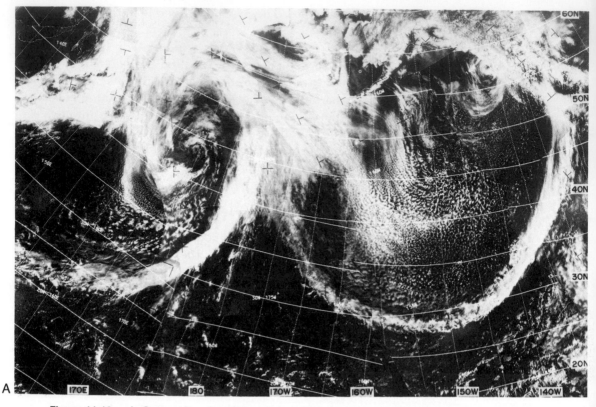

**Figure 11.10** *A. Composite of satellite photographs (April 22, 1969), showing cloud patterns associated with occluding wave cyclones over the Pacific Ocean. [Courtesy of National Environmental Satellite Center, NOAA.] B. Weather chart showing sea-level isobars and fronts at the time that the satellite photographs were taken.*

computers, much of this empiricism has been replaced in weather forecasting. However, because of the complexity of the equations and the inadequacy of the observations of the initial conditions, the numerical solutions are only approximate, and, particularly with respect to forecasting cyclogenesis, they are frequently inadequate. It is expected that improved observations, better mathematical models, and larger computers will lead to better forecasts in this regard.

Questions, Problems, and Projects for Chapter 11

1 Make graphs showing the variation of temperature with distance in Figure 11.1, (a) along an east-west line approximately dividing the map in two and (b) along a north-south line approximately following the Mississippi River. In the graphs mark the positions of any fronts that the lines intersect. Describe how the temperature changes in crossing the fronts and how it varies within the air masses on each side. Describe how the cloudiness and precipitation change in crossing the fronts and how they vary within the air masses on each side.

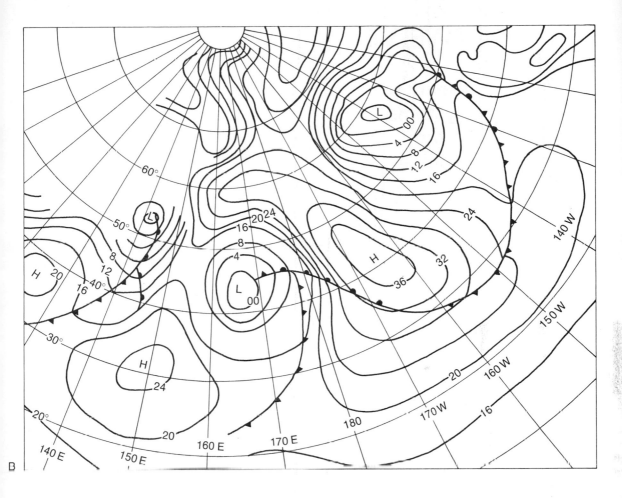

2 a Why do the pressure at the low center decrease and the winds around the low center get stronger as a wave cyclone occludes?

b Why do wave cyclones in middle latitudes usually move with a component toward the east?

3 a Why does the westerly component of the wind usually increase with height in the lower troposphere?

b Why are the strongest westerlies frequently concentrated in a narrow band or jet?

4 Why are thunderstorms more likely to be associated with cold fronts than with warm fronts?

# TWELVE

## Weather Systems in the Tropics

### 12.1  The atmospheric circulation at low latitudes

We have seen that at middle and high latitudes the temperature and state of weather undergo continuous variation, determined by the properties of one or another moving air mass and the passage of cyclones with their associated systems of cloud and precipitation. At low latitudes, between the two subtropical high-pressure belts, the variability is much less and except on infrequent occasions the circulation pattern on any day departs very little from the average for the season.

Near the equator the influence of the earth's rotation on horizontal motions is small. The horizontal component of the Coriolis force decreases with latitude and vanishes at the equator. The closer to the equator, the more the winds deviate from the geostrophic direction and speed, and the less useful the isobars are to summarize the wind field. Instead *streamlines* are used, which are lines having the property that at every point along them the tangent is in the direction of the wind.

Figure 12.1 shows the average isobars and streamlines over the earth between 50°N and 50°S latitude in January and July. The shaded areas are the regions of least variability of the wind, with wind direction within 45° of the average streamline more than 80 percent of the time. The shading is mostly in the trade winds equatorward of the subtropical HIGHS, showing

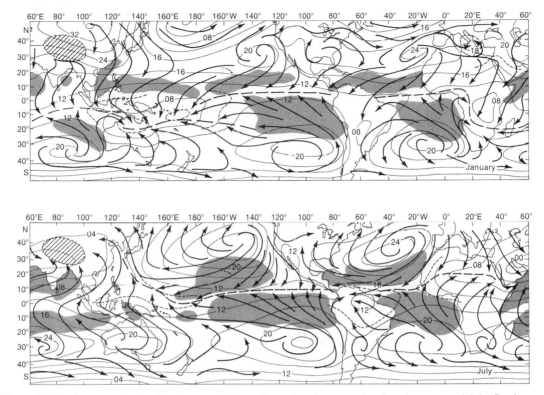

**Figure 12.1**  *Average sea-level isobars and streamlines at surface over tropics, January and July. Regions of steady winds, with wind direction within 45° of the average streamlines more than 80 percent of the time, are shaded.* [*From E. Palmén and C. W. Newton,* Atmospheric Circulation Systems *(Copyright 1969 by Academic Press).*]

that the trade winds are the steadiest features of the atmospheric circulation near the ground. Nevertheless, as we shall see, within this degree of steadiness the easterly trades are subject to significant fluctuations.

Notice that the streamlines cross the isobars at large angles near the equator. The convergence zone between the trade winds of the two hemispheres, known as the Intertropical Convergence Zone (ITC), is shown by heavy dashed lines. It is interrupted or displaced by the influence of continents. In general, the ITC migrates, being in the Northern Hemisphere in the northern summer and the Southern Hemisphere in the southern summer. However, over the Atlantic and eastern Pacific oceans it is north of the equator in January as well as in July. Over the Indian Ocean and the western Pacific the winds are dominated by the Asiatic monsoon, with continuous flow outward across the equator from the anticyclone over Asia in January, and inward toward the LOW over Asia from south of the equator in July. In January the northeast flow turns to northwest in crossing the equator; in July the southeast flow changes to southwest in crossing it.

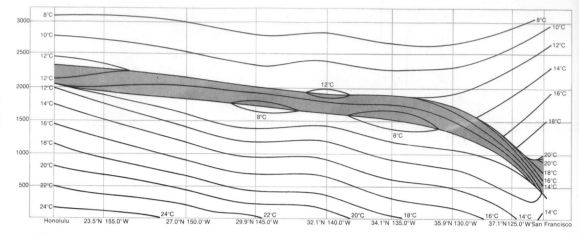

**Figure 12.2**   *Average vertical cross section from San Francisco to Honolulu in summer. Sloping solid lines are isotherms. The inversion layer is shaded. [From M. Neiburger, D. S. Johnson, and C. Chen. Studies of the Structure of the Atmosphere over the Eastern North Pacific Ocean in Summer, vol. 1:1 (1961). Published by the University of California Press; reprinted by permission of The Regents of the University of California.]*

## 12.2   Characteristics of the trade-wind region

Since the trade winds are directed equatorward, for the most part they carry air toward regions of warmer water. Thus the air near the sea surface is heated from below and has moisture added to it. In response to the instability thus produced, convective clouds — trade-wind cumulus — are characteristic of the region. At the same time, as discussed in Section 10.3, the flow around the subtropical anticyclone is divergent and an inversion is present that limits the height to which convective clouds can penetrate.

The trade-wind inversion is lowest at the eastern portions of the oceans, and slopes upward toward the west. It also slopes upward toward the ITC. Figure 12.2, which is an average vertical cross section from San Francisco to Honolulu in summer, illustrates the upward slope of the inversion as one follows a streamline southwestward. At the coast the inversion base is at 400 m, and the clouds can form only in the moist air below that height. Consequently, fog, stratus, or stratocumulus are present much of the time along the California coast. Farther southwest the inversion slopes upward, first rapidly, then less rapidly, and there is room for convective activity below it, permitting the typical trade cumulus to form. Over the eastern portion of the oceans the clouds still are not deep enough to produce rain, but farther west showers occur, particularly at times when the inversion is raised or dissipated by passing disturbances.

The motion of the air toward the west on both sides of the equator has a remarkable effect on the temperature of the sea surface. This is a consequence of the effect of the Coriolis force on the motion of the water due to the frictional stress exerted on it by the wind. The Coriolis force acts to propel the water to the right of the wind in the Northern Hemisphere and to the left in the Southern Hemisphere. The easterly winds on the two sides of the equator thus cause the surface water to move northward north of the equator and southward south of the equator. To replace the water that diverges at the equator, water must rise from the depths, where it is cold. The *upwelling* due to the frictional drag or stress of the easterly trades thus produces a belt of cold water along the equator over the eastern part of the oceans. Where this cold belt is present there is no heating of the air from below, and no convective cloudiness and precipitation are present. This part of the equatorial belt, contrary to expectation, is arid.

## 12.3   Waves in the easterlies

The existence of disturbances in the trade winds was first recognized in the 1930's by tracing the movement of pressure changes across the tropical North Atlantic Ocean. It was found that these fluctuations in pressure moved westward and were correlated with variations in the thickness of the moist layer below the inversion as shown in airplane soundings at the Virgin Islands. When more observations, both at sea level and aloft, became available the structure of the disturbances was shown to have the form of wave-shaped perturbations in the easterly flow somewhat similar to the waves in the westerlies aloft at high latitudes.

Whereas the instability that leads to the formation of waves in the westerlies is associated with baroclinicity, that is, the existence of horizontal temperature gradients, in the tropics the temperature varies very little. The circumstances under which instability occurs in easterly currents that are barotropic are still being investigated; at present we can recognize their presence but cannot forecast their formation.

Figure 12.3 is an example of a wave in the easterlies over the central Pacific. In addition to the isobars and symbols for surface winds, the figure contains dashed lines drawn through places with the same total water-vapor content in the vertical, and the winds at 12 km are shown by dashed symbols. The trough of the wave is shown by a heavy line; to the east of it the surface winds are from the east-southeast and the moisture content is high, corresponding to a thick, moist layer. To the west of the trough the low-level winds are mostly from the east-northeast, and the integrated vapor content is much less. In the upper troposphere the winds are from the west, as in middle latitudes.

Waves in the easterly trades move westward. In advance of the arrival of the trough of a wave in the Northern Hemisphere the winds are east-northeast and, corresponding to the relatively low vapor content, the cloudiness is less developed. With the passage of the trough the winds become east-southeast and the convective cloudiness becomes more active, with showers and thunderstorms. Figure 12.4 shows the vertical cross

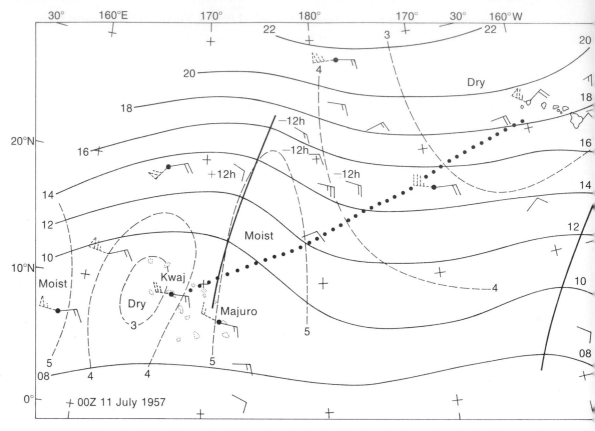

**Figure 12.3** *Sea-level isobars (solid lines) and winds (wind symbols) and vertically integrated water-vapor content (g/cm²) and winds at 12 km (dashed symbols) over central Pacific Ocean. [From Palmén and Newton (1969).]*

section along the dotted line in Figure 12.3. Ahead (to the west) of the trough, the clouds are only about 1 km deep, with low relative humidity above. Behind it is the region of deep convective clouds and showers.

Waves in the easterlies are generally about 2000 km long, and move with speeds of about 5 m s⁻¹. Thus it takes about four or five days for an entire wave to pass. Much longer waves, 5,000–10,000 km long, have been found in the upper troposphere and lower stratosphere in the tropics. They also travel westward and have periods of four to five days. They have no direct effects on the weather. The possibility that when they are superposed on a lower level disturbance they may tend to amplify it and in some cases to convert it into a cyclonic vortex has been suggested.

## 12.4 Tropical cyclones, hurricanes, and typhoons

Ordinarily, the weather in tropical regions is characterized by only small variations, with dry spells and showery spells in those areas affected by

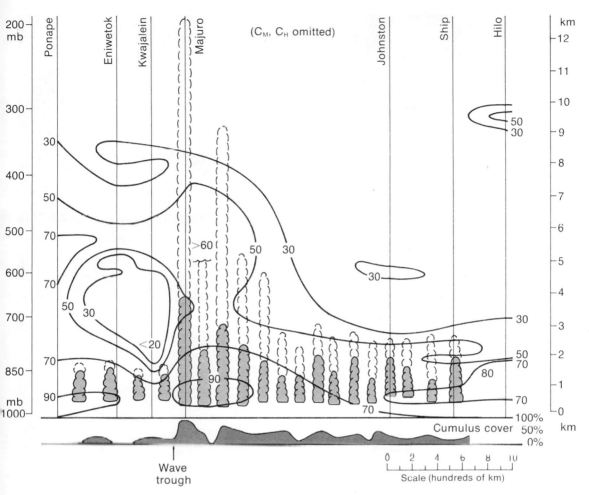

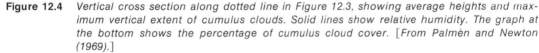

**Figure 12.4** *Vertical cross section along dotted line in Figure 12.3, showing average heights and maximum vertical extent of cumulus clouds. Solid lines show relative humidity. The graph at the bottom shows the percentage of cumulus cloud cover. [From Palmén and Newton (1969).]*

easterly waves, but mostly the weather is the same day after day. Occasionally, however, tropical disturbances develop into cyclones that may achieve such intensity as to cause the loss of many lives, blow down trees, and destroy or severely damage buildings. The intense tropical cyclones are known in various parts of the world by different names: *hurricanes* when they occur in the Atlantic or along the west coast of Mexico, *typhoons* in the western Pacific, *baguios* in the Philippines, *willy-willies* in Australia, and simply *cyclones* in the Indian Ocean, including the Bay of Bengal and the Arabian Sea. For convenience we shall refer to all of them as hurricanes.

Tropical cyclones are called hurricanes (or typhoons) only when their winds, in the part of the storm where they are strongest, exceed $32$ m s$^{-1}$. By international agreement, when they have winds between 17 and 32

**Figure 12.5** *View of Hurricane Gladys, taken by astronauts aboard Apollo 7 at 1531 GMT (1031 EST), October 17, 1968* [*Courtesy of NOAA.*]

m s$^{-1}$ they are called *tropical storms*, and when the winds in them are less than 17 m s$^{-1}$ they are called *tropical depressions*. The terminology of the U.S. National Weather Service includes an additional category, *tropical disturbances*, for incipient storms that have a slight cyclonic circulation at the surface but no more than one closed isobar on the weather map. Also, the U.S. National Weather Service uses slightly different values of the wind speed for distinguishing between the other categories. Tropical cyclones undergo development from disturbance to depression to storm to hurricane in a matter of a few days. They may continue as mature hurricanes for periods as long as two weeks or more, until they move onto land or out of tropical latitudes. When a storm moves from the tropics to higher latitudes, it usually maintains hurricane intensity but rapidly changes in structure to become an extratropical cyclone.

Fully mature hurricanes range in diameter from 100 km to more than 1500 km. The winds spiral inward toward the center, where the pressure is usually below 970 mb, with increasing speeds that reach values between 50 and 100 m s$^{-1}$ near the center. At the center itself, however, there is a region of light winds, about 5 m s$^{-1}$ or less, called the *eye of the storm*. Within the eye there is no rain or low cloud, middle clouds are scattered or broken, and cirrus and cirrostratus clouds are visible and partially clearing. Surrounding the eye is the eye wall, a circle of towering cumulonimbus clouds from which heavy rain falls, and spiralling inward to the eye wall from the edge of the storm are bands of cumulus and cumulonimbus with rain.

The spiral, banded structure is shown in Figure 12.5, which is a photograph of Hurricane Gladys taken October 17, 1968, by the astronauts on the Apollo 7 space flight. The photograph was taken looking southeast when the space vehicle was over the Gulf of Mexico at a height of 185 km and the storm was centered about 200 km west of Florida. The circular cap of cirrostratus near the storm's center is probably associated with the glaciation of the cumulonimbus clouds in the vicinity of the wall of the eye. The spreading of the cirrostratus prevents the structure near the center from being seen from above. Gladys had formed two days earlier in the Caribbean Sea in a region of widespread shower activity and moved northward about 5 m s$^{-1}$, becoming a hurricane shortly before crossing western Cuba. Winds exceeding 35 m s$^{-1}$ were observed in Cuba, and heavy flash floods with severe damage to crops and industrial installations were experienced. Continuing northward and north-northwestward, it crossed the Florida Keys and reached the position to the west of Florida where it was photographed from the spacecraft. It then curved toward the east and passed inland, causing damage to the Florida citrus crop and structural damage estimated to be about $6 million. Although the measured rainfall in some places was as high as 8 inches, flooding from rain was not serious. After crossing Florida it moved northeast along the coast, passing near Cape Hatteras, and gradually became extratropical.

The banded structure, together with the eye, enables easy identification of hurricanes on satellite photographs. Since hurricanes form over tropical oceans where surface stations are sparse and ships are infrequent, satellites provide an invaluable forecasting aid in detecting and tracking them. Figure 12.6 gives an example of a satellite photograph of a hurricane.

Tropical cyclones always have their beginnings in regions where the sea surface temperature over large areas is high. In Figure 12.7 the generalized tracks representative of tropical cyclones that reach hurricane intensity are shown. They all originate in areas where the sea surface temperature is higher than 27°C in the season of their occurrence. Furthermore, hurricanes do not form closer to the equator than about 5°N and 5°S, presumably because the Coriolis force at lower latitudes is too small to

**Figure 12.6** *Satellite photograph of Hurricane Beulah, taken at 1828 GMT (12:28 P.M. CST), September 18, 1967, over the Gulf of Mexico. The outlines of the United States, Mexico, and Cuba have been added in light dotted lines; the parallels of latitude and meridians of longitude, in heavy lines. [Courtesy of NOAA.]*

create the rotational circulation. The sea surface north of 5°N has its highest temperatures from June to November, and these are the months in which hurricanes develop most frequently, although occasionally hurricanes occur in May in the North Atlantic, and in the western North Pacific and the North Indian oceans tropical cyclones have occurred in every month of the year. Correspondingly, the period from December to March is the part of the year when the sea surface temperatures are highest and tropical storms are most frequent in the Southern Hemisphere. How-

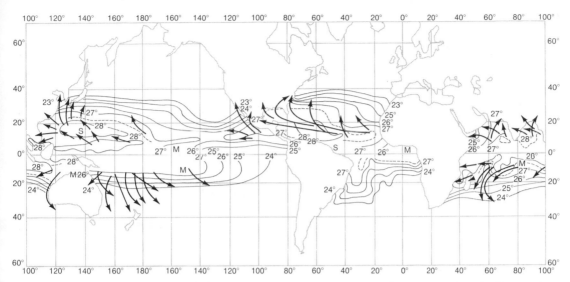

**Figure 12.7** *Principal tracks of tropical cyclones in relation to sea surface temperature. Isotherms show average sea surface temperature in September for the Northern Hemisphere and in March for the Southern Hemisphere. [From Palmén and Newton (1969).]*

ever, there have never been any hurricanes in the South Atlantic Ocean, presumably because the sea surface temperatures do not reach sufficiently high values there.

The warmth of the sea surface as a prerequisite for the formation of hurricanes is obviously related to the creation of air that is sufficiently unstable and moist to remain warmer than the surroundings, while rising saturation adiabatically to high levels. An adequate theoretical explanation of the formation of tropical cyclones has not yet been developed, that is, a theory that will make it possible to select from the frequent occurrences of high sea temperatures at appropriate latitudes the rare occasions when circumstances are right for the development of an intense cyclonic vortex. The large amounts of energy come from the latent heat released by the condensation in the rising air currents of the moisture that has evaporated from the warm sea surface. The air at the surface picks up heat and moisture as it moves inward toward the center. The added heat from the warm water compensates for the temperature decrease that would otherwise occur because of the decrease in pressure.

The general structure of a hurricane is shown schematically in Figure 12.8. The air is shown streaming toward the center at the surface. It gains rotational momentum due to the horizontal convergence, and rises both because of convergence and because of the hydrostatic instability that leads to intense convection, particularly near the eye wall. The upward velocity is greatest near the eye, and smaller farther out, but there is convergence and upward motion in the spiral cloud bands, with relative clearing in the zones between the cloud bands. In the eye there is sinking

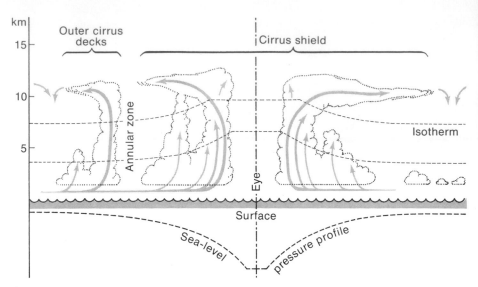

**Figure 12.8** *Schematic representation of a vertical section through a hurricane. The weight of the streamlines is intended to indicate the strength of the inflow and upward motion. Lower part of figure shows pressure variation at sea level.*

motion. At the eye wall the descending air tends to mix with the cloudy air around the eye. As a result of the adiabatic heating of the descending air, the eye is warmer than the surroundings right up to the tropopause.

At upper levels the rising air flows outward, and at the outskirts of the storm there is sinking motion. The hurricane is a warm-core cyclone, which must decrease in intensity with height and eventually become an anti-cyclone. Except near the eye, the outflow in the upper troposphere is anticyclonic.

The mature hurricane can be regarded as being composed of four parts: (1) an outer region in which there are cyclonic winds with higher speeds closer to the center and slight convection at low levels and flow outward from the center with cirrus or cirrostratus clouds at high levels; (2) a belt of squall lines, frequently in the form of spiral cloud bands, with winds reaching hurricane strength and heavy convective showers; (3) the ring-shaped inner region of extremely strong winds and heavy rains; and (4) the eye, inside a transition zone in which the wind drops almost to a calm and the precipitation ceases. Over the ocean the winds create tremendous waves, the tops of which are carried as heavy spray that merges with the rain "so that the mariner can scarcely tell where the ocean ends and the atmosphere begins."

As a hurricane approaches land it drives water ahead of it in the form of a "tidal wave" or storm surge that inundates coastal lowlands with very deep water. Frequently the storm surge causes more deaths and damage than the winds or flooding due to rain. Thus Hurricane Camille was accompanied by a 24-ft tide when it crossed the Louisiana-Mississippi

**Table 12.1**  Loss of Life and Property Damage from Some of
the Most Severe Hurricanes in the United States

| Property damage | | |
| --- | --- | --- |
| Hurricane | Year | Millions of dollars |
| Camille | 1969 | 1420.7 |
| Betsy | 1965 | 1420.5 |
| Diane | 1955 | 831.7 |
| Carol | 1954 | 461.0 |
| Donna | 1960 | 426.0 |
| Carla | 1961 | 408.2 |
| New England | 1938 | 306.0 |
| Hazel | 1954 | 251.6 |
| Dora | 1964 | 250.0 |
| Beulah | 1967 | 200.0 |
| Audrey | 1957 | 150.0 |
| Cleo | 1964 | 128.5 |
| Miami | 1926 | 112.0 |

| Loss of life | | |
| --- | --- | --- |
| Hurricane | Year | Deaths |
| Galveston | 1900 | 6000 |
| Louisiana | 1893 | 2000 |
| South Carolina | 1893 | 1000–2000 |
| Okeechobee | 1928 | 1836 |
| Keys and Texas | 1919 | 600–900 |
| Ga. and S.C. | 1881 | 700 |
| New England | 1938 | 600 |
| Keys | 1935 | 408 |
| Audrey | 1957 | 390 |
| Atlantic Coast | 1944 | 390 |
| Mississippi and Louisiana | 1909 | 350 |
| Galveston | 1915 | 275 |
| Mississippi and Louisiana | 1915 | 275 |
| Camille | 1969 | 256 |

coast the night of August 17–18, 1969. It is impossible to tell whether the 90 m s$^{-1}$ (200 mi/hr) winds or the tremendous hurricane wave contributed more to the toll of 256 deaths and almost $1.5 billion in property damage.

Table 12.1 gives the cost in lives and property due to hurricanes for a selection of severe storms of the past century. While the property damage in recent years has been great because of increased building of homes and industry in hurricane-prone regions, the loss of lives has been relatively small because of improved forecasting and warning-dissemination methods.

Tropical cyclones in the Bay of Bengal have been responsible for greater losses of life than any other natural disasters. Several times in the

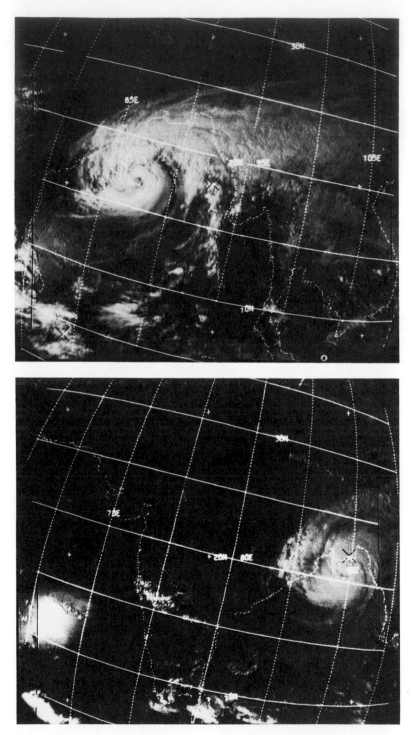

**Figure 12.9** *Bay of Bengal cyclone, November 11–12, 1970, photographed by ITOS 1 Satellite. [Courtesy of NOAA.]*

past three centuries cyclones have caused the inundation of the low land at the mouths of the Ganges, causing hundreds of thousands of deaths by drowning. The most recent case was the cyclone of November 12, 1970, in which the official death toll, confirmed by the number of burials, exceeded 200,000, and unofficial estimates indicate that as many as 500,000 people died. Figure 12.9 presents satellite photographs of that Bay of Bengal storm on two successive days. Among the previous occurrences, a storm in 1876 is reported to have caused a 40-ft tide that was responsible for 100,000 to 400,000 deaths, and on October 7, 1737, a storm with a similar tidal wave killed an estimated 250,000 people and destroyed 20,000 boats and ships.

It would seem clear that people should not live in low coastal areas subject to hurricanes. With increased understanding of meteorological conditions, and improved methods of forecasting them, we can expect the adjustment of human activities to take better advantage of the favorable aspects and avoid the unfavorable aspects of our atmospheric environment.

Questions, Problems, and Projects for Chapter 12

1   Discuss the similarities and differences between waves in the easterlies and waves in the westerlies.

2   Would you expect to find frontal waves at the zone of convergence between the trade winds of the two hemispheres (the Intertropical Convergence Zone, ITC)? Explain.

3   Why don't hurricanes ever occur right at the equator?

4   Why don't hurricanes develop in the Atlantic during the months of December to April?

5   Why do the winds in hurricanes decrease rapidly when the storms pass over land?

# THIRTEEN

## Weather
## Forecasting

### 13.1    Methods of weather prediction

In Section 1.3 we saw that we were able to forecast with some accuracy the 24–hour movement of a cyclone from its past movement and to estimate the weather changes that would occur in the area over which it passed by consideration of the weather conditions around the low as shown on the synoptic weather map. With additions and refinements, this empirical method of forecasting was used for the daily forecasts issued by weather services throughout the world until late in the 1950's. Based on the synoptic weather map, it is called *synoptic weather forecasting.*

In the past two decades a more rigorous method, which attempts to apply the laws of physics, has been developing. Since the solution of the equations expressing those laws for atmospheric motions requires the use of numerical computations carried out on high-speed electronic digital computers, the method is generally referred to as *numerical weather prediction* (NWP), although it would be more proper to call it *dynamic weather prediction* to distinguish it from other methods that include numerical procedures.

*Statistical forecasting* is another method in which numerical procedures are used, first, to analyze past data to establish the predictive relationships, and then to utilize the current data to obtain the prediction. A particular kind of statistical relationship that is sometimes sought is *periodicities* or *cycles*. If it could be established that the weather repeated itself at specific intervals, the problem of forecasting would be solved. A related approach is the *analogue method*, in which past weather situations similar to the current one are sought. If a situation identical to the present weather could be found, the future weather should be a repetition of the weather that occurred subsequent to the analogue. In that case the weather would have a periodicity equal to the time between the analogue and the present. In actuality, no well-marked periods other than the diurnal and annual have been found, and while past weather situations can be found in which some of the features resemble the present fairly closely, there are always enough differences to insure that the development will be different.

The forecasting procedure used at present involves a combination of the numerical, synoptic, and statistical methods. The general large-scale flow patterns are computed at weather centrals using NWP. The results are modified somewhat by applying corrections to allow for errors that statistical analysis of past computer forecasts have shown. They may be further adjusted in accord with the forecaster's judgment of the reasonableness of the computer output, based on his synoptic experience with similar situations. The flow patterns computed include estimates of the vertical velocity, which give a first indication of areas of cloudiness and precipitation. For details of the occurrence of precipitation and forecasts of temperature, statistical correlations are frequently used. In all of this, allowance is made for the two well-defined periodicities of the atmosphere, the diurnal and annual cycles.

The forecasts of the large-scale flow patterns are made at national weather centrals. The more detailed forecasts of the weather for particular localities and for particular purposes are made at regional and local weather stations. The forecasts prepared by the weather centrals are transmitted to the regional and local stations by telefacsimile in the form of prognostic charts as well as by teletype or radio in verbal and symbolic messages. The regional and local forecasters adapt to their localities the material sent by the centrals, utilizing their synoptic experience and statistical correlations, as well as local indications.

## 13.2  Predictability of the weather. Accuracy of weather forecasts

With all these methods available the question arises, Why are the daily weather forecasts that are broadcast on television and radio frequently in error? The same question might also arise if we compare prediction of the weather with the prediction of eclipses of the sun and the moon and the movement of planets among the fixed stars. If astronomers can forecast to the minute the occurrence of an eclipse many years in advance, why cannot meteorologists forecast with similar accuracy the occurrence of precipitation just one day in advance?

The answer lies in the complexity of the physical systems involved. For the prediction of eclipses, the motions of the sun, moon, and earth can be treated as a problem involving three bodies, each with its mass concentrated at a point. This three-body problem is not simple, but it is infinitely less complex than keeping track of the movements of the innumerable parcels of air with which one must be concerned in forecasting the weather. The problems of celestial mechanics reduce to the dynamics of a small number of isolated mass points moving in response to the forces they exert on each other. The problems of meteorology are those of the dynamics of a thermally active fluid, and fluid dynamics is innately more difficult than particle dynamics.

If the problem were only to forecast the movement of the high and low centers and if they obeyed simple laws of interaction, it might be no more difficult than the eclipse problem. But in addition to the movement of the centers at sea level it is necessary to forecast their intensity there, their position and intensity at all other levels, the distribution of pressure in between, the distribution of temperature at various heights, the variations of humidity, the occurrence of clouds and precipitation, and the wind velocity. The laws governing the complex interactions of these factors are only partially understood. To the extent that they are understood, they can be represented by mathematical equations, but it turns out that the solution of the equations in their full complexity would require computers of much greater capacity and speed than the largest so far available.

Another limitation on forecast accuracy is due to the fact that the changes in weather conditions at any one place are influenced by the conditions over a large area surrounding it, the size of the area increasing with the length of time for which the forecast is being made. The forecast of tomorrow's weather requires a complete delineation of the present pressure, temperature, humidity, wind, and so forth, throughout the troposphere and lower stratosphere over a large part of a hemisphere. Observations are made only at a discrete number of stations, sometimes so far apart that important weather features may lie between and be missed. At the surface over land, at least in the more advanced countries, a fairly dense network of stations report the conditions, but over the sea and in less-developed countries the observations are sparse. There are fewer upper-air observations even in the areas where surface observations are adequate, and over much of the earth observations of conditions aloft are absent. If we do not know the present weather sufficiently well, how can we expect to forecast it for the future?

In spite of these difficulties and deficiencies weather forecasts at a useful level of accuracy have been made for many years. The beginnings of weather forecasting coincided with the development of the electric telegraph. Previously, there had been precursors of the synoptic weather map, which were based on past data collected by mail. The first of these was a series of charts showing winds and departure of pressure from average for each day during the year 1783. These charts were prepared in 1820 by Heinrich W. Brandes (1777–1834), who may be regarded as the first German meteorologist. The availability of the observations was due to the foresight of Antoine L. Lavoisier (1743–1794), the famous French chemist. In about 1780, Lavoisier proposed setting up strictly comparable instruments for taking meteorological measurements at a large number of places in France, Europe, and even the entire world. Acting on his own proposal, he distributed several barometers to observatories in Europe, enabling Brandes subsequently to collect the data and prepare his charts. Shortly afterward, James P. Espy (1785–1860), an American meteorologist, organized an observational network in the United States and drew daily weather maps on which the winds, pressure, and precipitation observed at the stations of the network were summarized. From these early charts prepared by Brandes and Espy came the first demonstration that the low pressures associated with bad weather occur in areas that move regularly, usually from west to east. When the telegraph enabled the collection of observations on a current basis, the possibility of using this concept for weather forecasting became a reality.

Beginning in the 1860's, meteorological services in various countries of Europe were established to prepare daily weather maps and issue forecasts based on them. While telegraphic collection of reports and transmission of weather indications in the United States began in 1849 under the aegis of Joseph Henry (1797–1878), the first secretary and director of the Smithsonian Institution, a national weather service was not established and regular forecasting in the United States was not begun until 1870. Initially, the Weather Bureau was part of the Signal Service of the U.S. Army, then successively of the Department of Agriculture and the Department of Commerce. Now called the National Weather Service, it is a branch of the National Oceanic and Atmospheric Administration in the Department of Commerce. Whereas the attention of the public now seems frequently to be focused on the occasions when the forecasts are wrong, in the early years the successful prediction of changes in something so variable as the weather was so surprising that the misses were ignored or accepted, and, increasingly, reliance was placed on the forecasts by nautical interests, farmers, shippers, and the general public.

How accurate are weather forecasts? The answer depends on the particular weather feature being forecast and the length of the period and the locality for which the forecast is made. Wind direction and speed and temperature are among the elements for which the forecaster can claim greatest success. The occurrence of cloudiness and precipitation is usually forecast fairly well, although the timing may be off. Forecasts of amounts of precipitation are subject to larger errors. The optimum forecasts are those covering periods of 6 to 24 hours. Forecasts covering periods of 2 or 3 days are fairly accurate, although their accuracy decreases with each day. Beyond 72 hours the best that can usually be done over and above climatic expectation is to provide a general indication, although in some situations the conditions are sufficiently stable or slow-moving that specific forecasts for periods as long as a week can be made.

The length of time for which we can expect to be able to forecast the sequence of weather can be estimated by the methods of NWP by carrying out computations based on two initial weather situations that differ by an amount corresponding to the uncertainty or expected error of the observations, and seeing how rapidly the differences increase. If the forecasts based on the two initial conditions do not diverge, the expectation would be that observational errors would not affect the accuracy of the prediction, and the length of time for which the forecast could be made would depend entirely on the validity of the forecasting method. If, however, the forecasts progressively become more and more different, there would be no way of deciding which would correspond to what would really take place in the atmosphere, and the time it took for them to differ by an amount equal to the acceptable error in a forecast would be the longest time for which a forecast would have any validity. Tests using this procedure have indicated that good forecasts should be possible for periods as long as two weeks, but not for longer periods. Actually, the procedure depends on the degree to which the NWP computations correspond to the way in which the atmosphere behaves, so the conclusion might be changed if improved NWP methods are developed.

There have been many attempts to measure the accuracy of weather forecasts. A fundamental difficulty is the problem of deciding whether a forecast is right or wrong, and if wrong, how wrong. If the forecast says it will rain today in Kansas, and it rains in the western part of the state but not in the eastern part, is the forecast right, wrong, or one-half right? If the forecast says the minimum temperature tomorrow morning will be 40°F, and it turns out to be 41°F, should one say the forecast is as wrong as it would be if it were 50°F? The National Weather Service expresses the accuracy of forecasts as a percentage only for the occurrence of precipitation. For the United States as a whole the forecasts of precipitation average about 80 to 85 percent right. The average error in the temperature forecasts is about 4°F. To rate the success of the forecasts of ceiling and visibility in the forecasts for airways terminals, a point system is used. Scores that estimate the degree to which the temperature and precipitation forecasts are better than reliance on climatology or persistence are used in evaluating the success of the forecasting methods. In addition, statistical measures are computed of the correspondence of the isobaric contours on the prognostic maps prepared in numerical weather prediction to the observed maps. The main purpose of these evaluations is to spur the forecasters to greater efforts and to judge the improvement, if any, in the forecasts when new methods are introduced and when gaps in the observational network are filled.

These measurements show that the forecasts are considerably better than climatological averages, pure chance, or persistence, although they are far from perfect, and that there has been some improvement in recent years as the empirical methods have been replaced by the NWP and as the observed data has become more plentiful. The improvement has been greatest in the forecasting of the general pressure pattern aloft. Based on the pressure pattern, the wind forecasts for aviation are usually very close to the observed. The procedures used in deducing the changes in temperature and the occurrence and amount of precipitation from the forecast of the pressure and wind distribution have not improved as much.

### 13.3   Analysis of the present weather

The first step in forecasting, whether by empirical methods or by NWP, is to determine the present weather situation and to represent it in a fashion that can be comprehended by the human forecaster or used in the machine computation. This process takes place in three parts, (1) observation, (2) transmission and collection, and (3) plotting and analysis.

As discussed in Chapter 1, observations are characterized as surface and upper air. The observations that are made for use in forecasting are called *synoptic observations.* They are made simultaneously four times a day (0000, 0600, 1200, and 1800 GMT) at all stations and are standardized by international agreement through the World Meteorological Organization, a specialized agency of the United Nations, with more than 130 member countries. The regulations of the WMO insure that the instruments and the procedures used result in comparable accuracy of the observations throughout the world.

Surface weather observations are made on a regular schedule at more than 7000 weather stations over the earth and aboard many hundreds of naval, merchant and passenger ships as they ply the oceans. Even with this tremendous number of stations there are large land areas, particularly in tropical and polar regions, and tremendous portions of the oceans, away from the principal shipping lanes, where there are few observations.

The routine surface observations include the pressure, temperature, dew point temperature, cloudiness (amount, type, and heights), wind direction and speed, current weather, pressure tendency, and amount of precipitation since the last report. These observations are transmitted by radio, telegraph, or teletype every six hours in a standard code, which consists of five-digit "words." The form of the transmission of an observation and a sample message are as follows:

| IIIii | N dd ff | VV ww W | PPP TT | $N_h$ $C_L$ h $C_M$ $C_H$ |
|-------|---------|---------|--------|---------------------------|
|       |         |         | $T_d T_d$ a pp | 7 RR $R_t$ s |
| 72405 | 83220   | 12706   | 14701  | 67292 |
|       |         |         | 00228  | 74541 |

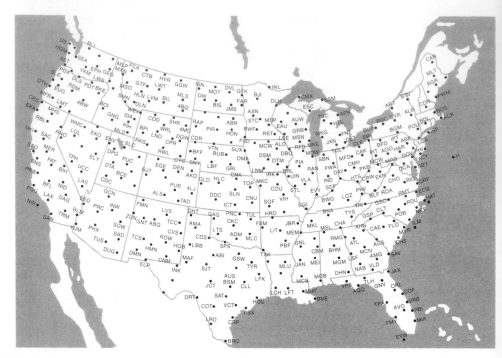

**Figure 13.1** *Surface observing stations in the United States.*

The first five-digit word gives the station identification number (Washington, D.C., in the sample message), and the remainder represents the weather data according to the symbols and tables given in Appendix D.

Figure 13.1 shows the locations of the stations in the contiguous United States (i.e., all the states excepting Alaska and Hawaii) that take surface synoptic observations.

The conditions aloft in the troposphere and lower stratosphere are observed principally with radiosondes, rawinsondes, and pilot balloons. The radiosonde observations (RAOBS) give the temperature, pressure, and humidity at various levels. In addition, the rawinsondes (RAWINS) also give the wind direction and speed, as determined by tracking the radiosonde transmitter with direction-finding equipment. The pilot balloon observations (PIBALS) are upper wind measurements made by tracking small balloons visually, using theodolites (small telescopes similar to surveyors' transits, but with the eyepiece along the axis of vertical rotation). For accurate wind measurements two theodolites (and thus two observers) separated by a known distance are required, but good approximate values are obtained by the single-theodolite method, which assumes a known ascension rate of the balloon. The data from RAOBS and RAWINS are reported for specified "mandatory" pressures—1000, 850, 700, 500, 400, 300, 200 mb, etc.—as well as at "significant" levels that enable reproduction of the details of the sounding. Like the surface observations,

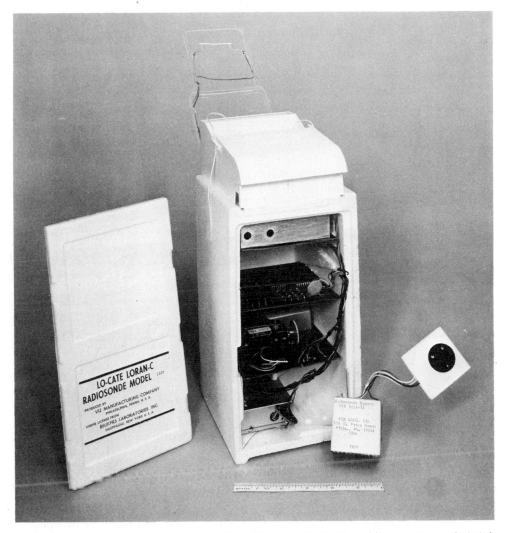

**Figure 13.2**   *Radiosonde for use in rawin observations. This recently developed instrument, carried aloft by a balloon, measures and transmits the temperature, humidity, and pressure as the balloon rises with greater accuracy than earlier models and enables determination of the upper-air winds by tracking, using navigational aid transmissions. The temperature is sensed with a thermistor, a ceramic rod whose electrical resistance varies with the temperature. Similarly, the humidity is measured by the electrical resistance of a slide that is coated with a carbon film. The pressure is measured by an aneroid baroswitch with 180 contacts, or points at which the pressure is accurately determined. For wind determination, the transmissions from Loran C or Omega navigational aid stations are received by the radiosonde and retransmitted to the ground station. Automatic comparison of the navigational aid signals received directly at ground stations and those received after being retransmitted from the balloon enables the position of the balloon to be determined, and from successive positions the wind speed and direction is derived. [Courtesy of Beukers Laboratories, Inc. and Viz Manufacturing Co.]*

**Figure 13.3** *Radiosonde stations in the United States. [Courtesy of NOAA.]*

RAOBS and the wind soundings are transmitted in a numerical code consisting of groups of five-digit words.

There are approximately 500 radiosonde or rawinsonde stations throughout the world, the densest network being in Europe. About 100 are taken at U.S. National Weather Service stations in the United States and on Caribbean and Pacific Islands, and about 50 more at other stations in the Western Hemisphere, including 7 at ship stations—5 in the Atlantic and 2 in the Pacific. The ship stations are maintained at specific locations. Coast Guard vessels carrying National Weather Service observers and equipment occupy the positions for about 21 days at a time, after which they are replaced by other vessels.

The RAOBS are taken twice daily, at noon and midnight Greenwich time. PIBALS are usually taken four times daily. In places where RAWINS provide wind data in conjunction with the RAOBS, the winds at the other two synoptic times are obtained using PIBALS.

Figure 13.3 shows the radiosonde stations in the contiguous United States. All of these stations take RAWIN observations.

In addition to the regular synoptic observations, there are a variety of observations that are made to supplement them or for special purposes. Among these are the hourly airways observations that are used to provide more frequent and more detailed information for pilots of aircraft, the pilot reports of conditions experienced by planes in flight (PIREPS or

AIREPS), and the radar and sferics observations that enable determination of the location and progress of precipitation areas and thunderstorms. The radar echoes show the position and intensity of storms from which precipitation is falling. Sferics is a system of determining the distance and direction of lightning strokes.

The satellite photographs provide additional information to that obtained from the observational network, both by corroborating and completing the data on cloud distribution in areas of dense conventional observations and by filling in the gaps where the conventional observations are sparse.

To determine the location and characteristics of hurricanes and typhoons specially equipped weather reconnaisance planes fly into the storms at various levels.

The various synoptic observations are sent by the observing stations to national and regional data centers. Collections of them are transmitted by these centers to the three World Meteorological Centers at Washington, Moscow, and Melbourne, Australia. In turn, the world centers and regional centers redistribute the collections as needed to the various forecast centers. In addition to the raw (unprocessed) observational data, the centers transmit, by facsimile, synoptic charts on which the present situation has been summarized and prognostic charts showing the expected developments. The forecast centers usually also prepare summary charts of their own, for more detailed representation of conditions in the area for which they forecast. Special and supplemental observations are also distributed to forecast centers.

The purpose of the synoptic charts, whether prepared at the weather centrals or at forecast stations, is to provide the forecaster with a three-dimensional picture of the current weather situation in a comprehensible fashion. Clearly, the forecaster cannot hold in his mind the hundreds or thousands of individual observations of conditions that may influence the weather over the area for which he is forecasting. By representing the observations on maps and other charts in an appropriate fashion, the separate data are brought together into patterns that are readily comprehended, such as high- and low-pressure centers, air masses, fronts, wave cyclones, waves in the westerlies, and jet streams. Using the physical principles that govern atmospheric processes, like those that have been discussed in the previous chapters, the forecaster is able to reason about the future behavior of these various features.

Preparation of the charts begins with the plotting of the observational data. The model for plotting the observations on the surface synoptic map was presented in Figure 1.15 and in Appendix D, and was discussed in Section 1.3. Once the data at all stations have been plotted, the map analyst proceeds to locate the fronts and to draw the isobars that summarize the pressure field. In placing the fronts, the principle of continuity is used.

**Figure 13.4**　*Model for plotting rawinsonde observations and reports from aircraft on upper-air charts.*

*HHH*　*Height of constant pressure surface, in meters with thousands digit omitted at 850 mb and in decameters at 500 mb.*

*TT*　*Temperature (°C).*

$D_nD_n$　*Dewpoint temperature depression (°C).*

*d*　*Wind direction in tens of degrees with hundreds digit omitted.*

*Wind direction is plotted to 36 compass points. Flag denotes 50 knots, barb denotes 10 knots, and half-barb denotes 5 knots. Circle is blacked in when dewpoint depression is 5°C or less.*

*GGZ*　*Time of report GMT to nearest hour.*

*hh*　*Altitude of aircraft in thousands of feet.*

Since fronts are the boundaries between air masses, continuity requires that the movement of the fronts from their position on the previous map be consistent with the movement of the air masses on either side. The characteristics of the fronts that help in identifying their position were listed in Section 11.2. For the times when RAOBS are available, the surface analysis is coordinated with the upper-air analysis. The upper-air data, particularly the isotherms at 850 mb and the isopleths representing the thickness of the layer between 1000 and 500 mb, help to locate the positions where significant differences in air-mass temperatures occur. Isobars are drawn in accord with the observed pressures, areas of current precipitation are shown by shading (colored green on manuscript maps), and special symbols are entered for such phenomena as squall lines and areas of severe thunderstorms.

The upper-level maps are drawn for some or all of the standard pressure values—850, 700, 500, 400, 300, 200, and 100 mb. In the U.S. National Meteorological Center these charts are all plotted and analyzed by the computer. The data are plotted by the machine in accord with the models shown in Figure 13.4. Contours showing the height of the pressure surface are drawn at 60-meter intervals, and the centers of high and low are labeled H and L, as for the high- and low-pressure centers on the sea-level map. It will be remembered that contours of isobaric surfaces are identical with isobars at a constant level, and have a similar relationship to the wind. Isotherms are drawn as dashed lines, at intervals of 5°C. Isotherms on a constant-pressure surface are also lines of equal potential temperature.

For the computer-produced charts, the observational data are fed directly from teletypewriter circuits into a computer that has been programmed to decode the messages, check the data for accuracy and consistency, and list the reports in an organized sequence. The computer then evaluates the height and temperature at fixed grid points (see Figure 13.5) on the

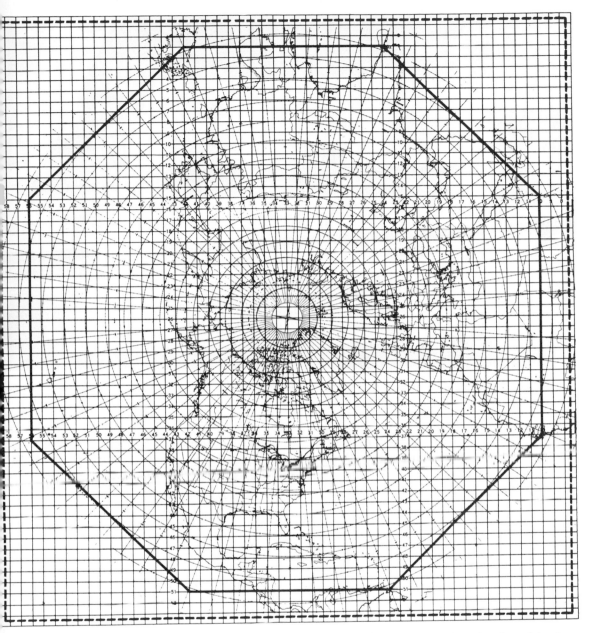

**Figure 13.5**  *Grid for numerical weather analysis and forecasting. The data are evaluated at the points of intersection of the horizontal and vertical lines over the map of the Northern Hemisphere. [From F. G. Shuman and J. B. Hovermale, "An Operational Six-Layer Primitive Equation Model," J. Appl. Meteorol. 7(4) (1968):528.]*

basis of the values at the locations of observations, and height contours and isotherms are drawn either by a printout procedure or by a "curve follower." Figure 13.6 shows a chart being drawn by a curve follower. In addition to the contours and isotherms, isotachs—lines of constant wind speed—are shown as dotted lines on these charts.

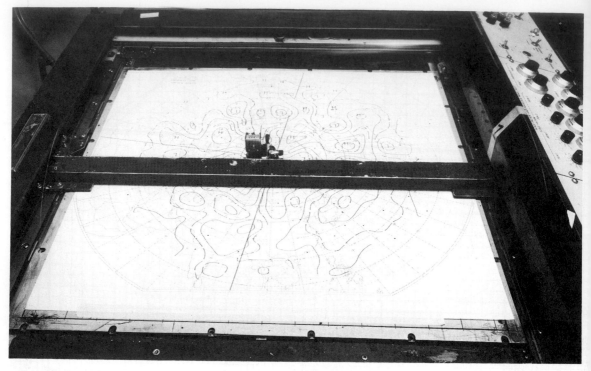

**Figure 13.6** *Curve follower for automatically drawing weather maps from computer-produced data. [Courtesy of NOAA.]*

A number of other upper-air charts are prepared at the National Meteorological Center and transmitted over the facsimile circuits to the forecasters throughout the nation. These show various derived quantities, such as the 1000- to 500-mb thickness referred to earlier and the 24-hour change in 500-mb heights.

## 13.4 Numerical weather prediction

Numerical (or dynamic) weather prediction is based on the assumption that weather phenomena obey physical laws expressed in equations that can be solved in the required detail and with sufficient accuracy. As has already been indicated, this is only partially true because of the complexity of the atmospheric system and because of the inadequacy of the observations defining the initial state. It would be hopeless to attempt to predict the motions and properties at all scales from global circulations down to the smallest eddies. However, it turns out that by appropriate treatment of the equations they can be made to represent only the large-scale motions. The solutions of these "filtered" equations can be carried out by using high-speed computers in such a fashion that the main, larger-scale features of flow patterns, such as cyclones, are forecast with considerable accuracy for three or more days in advance.

**Figure 13.7** *Communications equipment at the National Meteorological Center. The facsimile machines in the right foreground are used to transmit analyzed maps and prognostic charts of various kinds to stations throughout the world. [Courtesy of NOAA.]*

As an indication of the procedure, consider equation (5.1), which may be written

$$\Delta M\mathbf{v} = (\mathbf{F}_1 + \mathbf{F}_2 + \ldots)\,\Delta t \tag{5.1a}$$

This equation gives the change in momentum, $\Delta M\mathbf{v}$, of an air parcel of mass $M$ when forces $\mathbf{F}_1$, $\mathbf{F}_2$, and so forth, act on it for a short interval of time $\Delta t$. As shown by equations (5.4) and (6.1), the forces acting depend on the present values of the pressure gradient, the density (and thus the pressure and temperature), and the velocity. If we know these quantities at the present time, which we represent by $t_0$, we can use the equation to compute the velocity at time $t_1 = t_0 + \Delta t$. Analogous equations give the change in temperature, pressure, and so forth. With the values we have obtained for time $t_1$ we can repeat the procedure to obtain the velocity, pressure, and temperature at time $t_2 = t_1 + \Delta t = t_0 + 2\Delta t$. By repeated iterations we could find the distribution of these quantities for all parcels of air at any time in the future.

In addition to the magnitude of the task if we were to attempt to divide the atmosphere into small enough parcels to show all the small-scale motions and to keep track of the position and motion of each, there are computational problems that arise when the time scale, as indicated by the choice of $\Delta t$, and the distance scale [for instance, $\Delta n$ in equation (5.4)] are not properly related. Progressively, during the period since the first application of an electronic digital computer to the weather prediction problem in the late 1940's, these problems have been overcome. Methods have been developed that filter out the effects of small-scale ("sub-grid") motions and eliminate the possibility of explosive accumulation of computational errors. From the computational standpoint, the forecast could be extended for several months without "blowing up." It appears that the present limitation of accuracy to about three days is due to deficiencies in the equations specifying the physical processes and in the observations specifying the initial state of the atmosphere.

The early numerical weather predictions gave the flow patterns over a limited portion of the globe at a single level representative of the middle of the troposphere, which was taken to be the 500-mb pressure surface. By the present NWP method, solutions representing the flow in six layers over almost the entire Northern Hemisphere are derived.

Figure 13.5 shows the layout of points at which the data representing the initial state of the atmosphere are used in the computation and the predicted results are computed. The data are interpolated for the various layers at all the intersections of horizontal and vertical lines on the diagram from the analyzed constant-pressure charts.

An example of the results of an NWP prediction is shown in Figures 13.8 to 13.11. The initial pressure field at sea level and contours of the 500-mb constant-pressure surface are shown in Figure 13.8. The conditions forecast for these levels 24 hours later and the actual conditions observed are shown in Figure 13.9. In general, the movement of the main features — the troughs and ridges at 500 mb and the high and low centers at sea level — was predicted accurately, and some of the changes in intensity were forecast correctly, but the intensification of the low center that was approaching the west coast of North America was missed. In the 48-hour forecast, Figure 13.10, the differences between the predicted and observed are seen to be greater, particularly over the United States, where the increase in amplitude of the ridge at 500 mb led to winds that were stronger and had larger meridional components than are indicated by the forecast. However, even at 72 hours (Figure 13.11), many of the features of the forecast charts, for instance, the low over western Canada and the high over eastern Canada on the sea-level map, are in approximately the observed positions.

The NWP output includes predictions of the vertical velocity and the humidity distribution, from which quantitative precipitation forecasts can be made. Figure 13.12 shows the total precipitation that was predicted to fall between 12 and 24 hours after the initial time, compared with the observed amounts. While the agreement is far from perfect, the areas for which precipitation was forecast are generally correct, and the amounts are of the right order.

While the NWP gives estimates of the flow pattern that are better and can be made for longer periods than those made by the subjective and empirical methods in use previously, forecasts of temperature and precip-

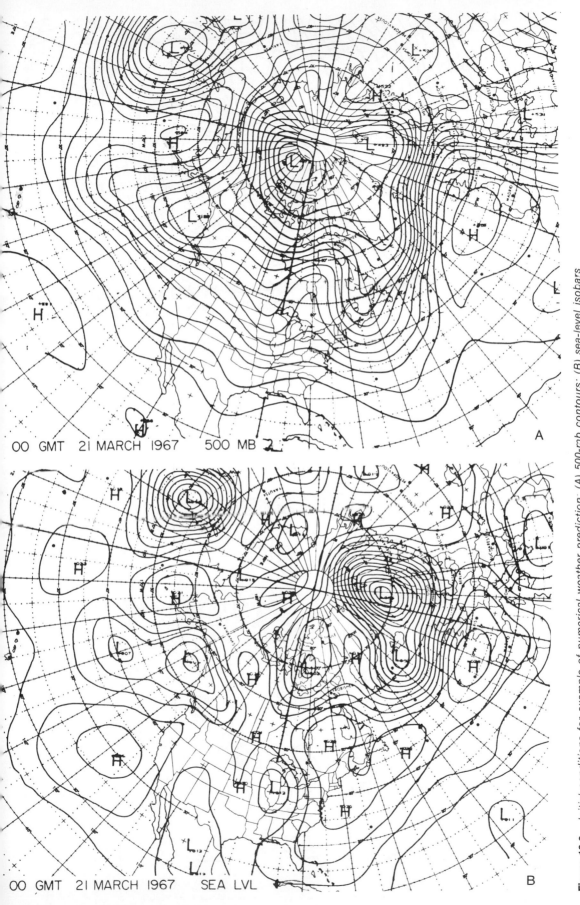

OO GMT  21 MARCH 1967    500 MB                                    A

OO GMT  21 MARCH 1967    SEA LVL                                   B

**Figure 13.8**  *Initial conditions for example of numerical weather prediction: (A) 500-mb contours; (B) sea-level isobars based on observations made at 0000 GMT, March 21, 1967. [From Shuman and Hovermale (1968).]*

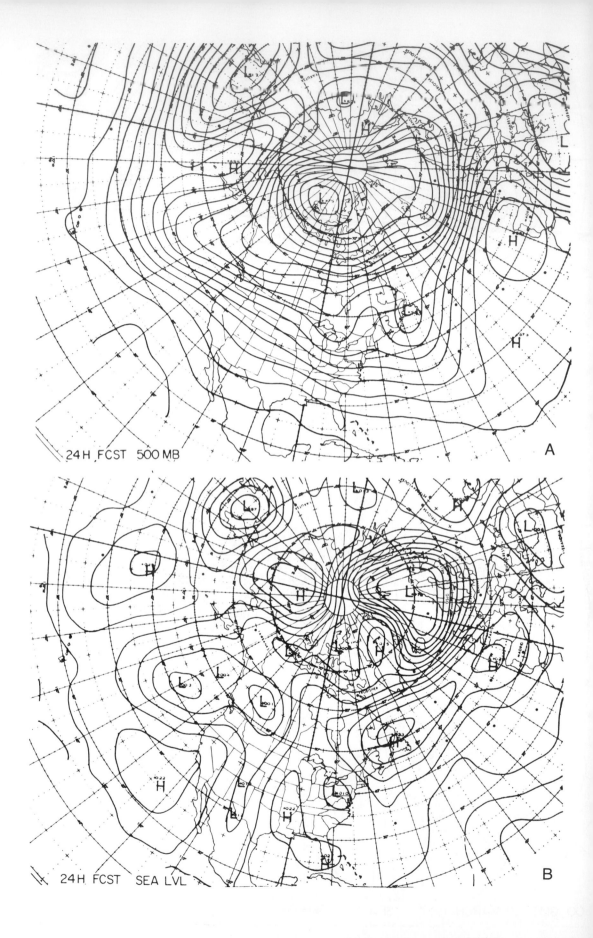

24H FCST 500 MB

A

24H FCST SEA LVL

B

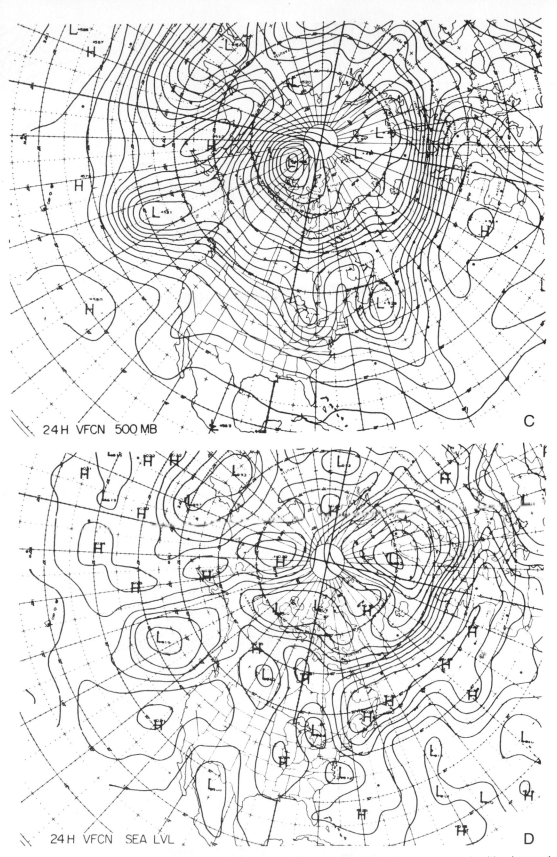

24 H VFCN 500 MB                                                C

24 H VFCN SEA LVL                                              D

**Figure 13.9**  *Numerically computed prediction for 24 hours after initial time, compared with observed situation at that time: (A and B) forecast of 500-mb contours and sea-level isobars for 0000 GMT, March 22, 1967; (C and D) corresponding observed 500-mb contours and sea-level isobars.* [*From Shuman and Hovermale (1968).*]

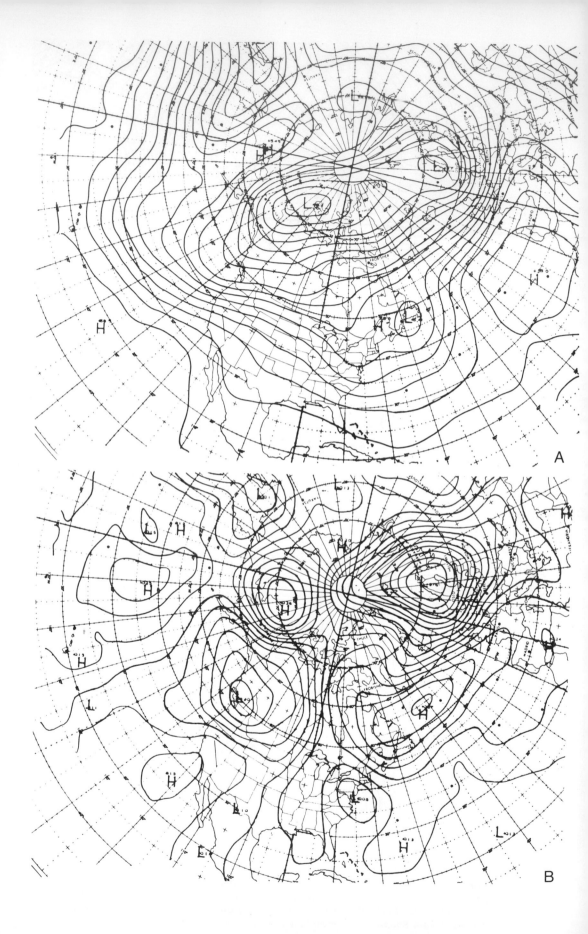

A

B

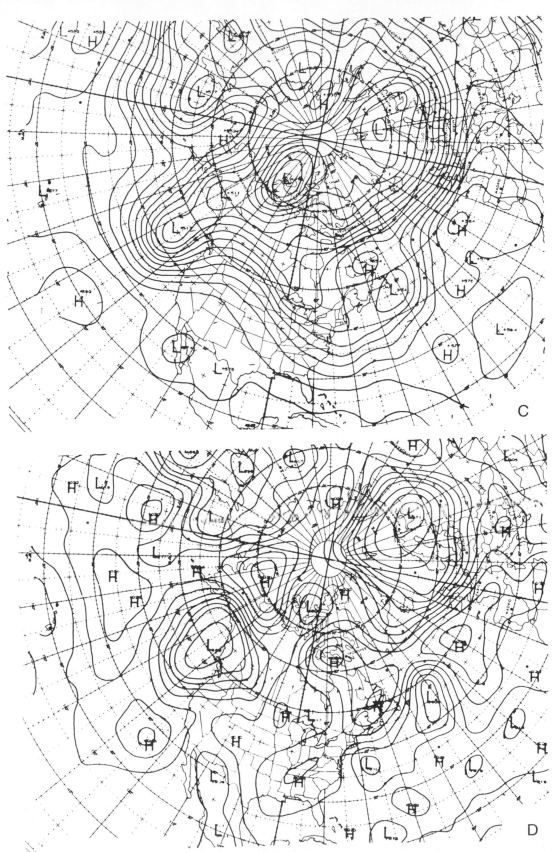

**Figure 13.10** *Numerically computed prediction for 48 hours after initial time, compared with observed situation at that time: (A and B) forecast of 500-mb contours and sea-level isobars for 0000 GMT, March 23, 1967; (C and D) corresponding observed 500-mb contours and sea-level isobars. [From Shuman and Hovermale (1968).]*

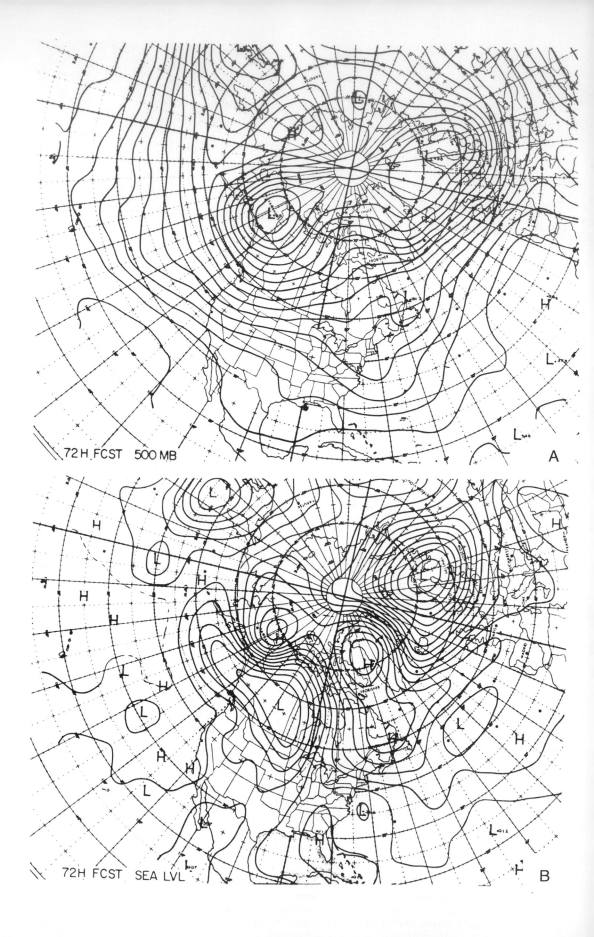

72H FCST 500 MB

A

72H FCST SEA LVL

B

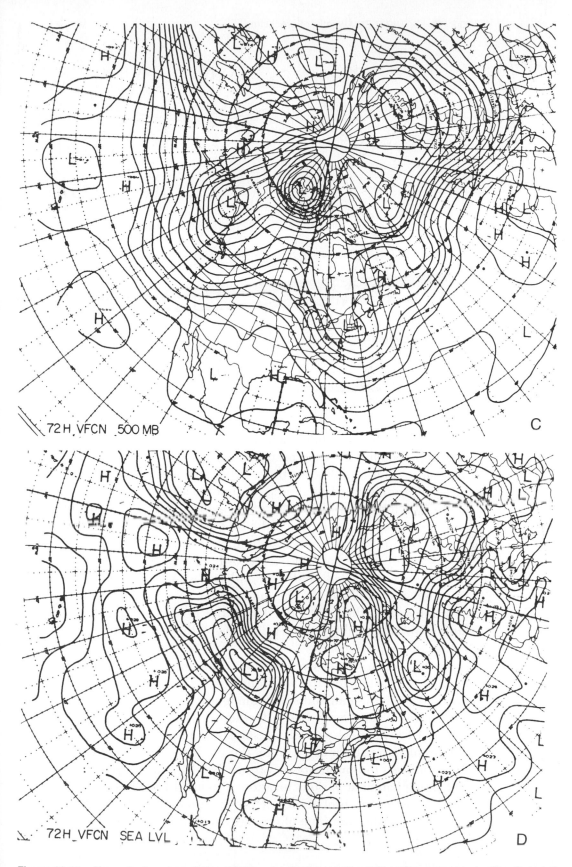

**Figure 13.11** *Numerically computed prediction for 72 hours after initial time, compared with observed situation at that time: (A and B) forecast of 500-mb contours and sea-level isobars for 0000 GMT, March 24, 1967; (C and D) corresponding observed 500-mb contours and sea-level isobars. [From Shuman and Hovermale (1968).]*

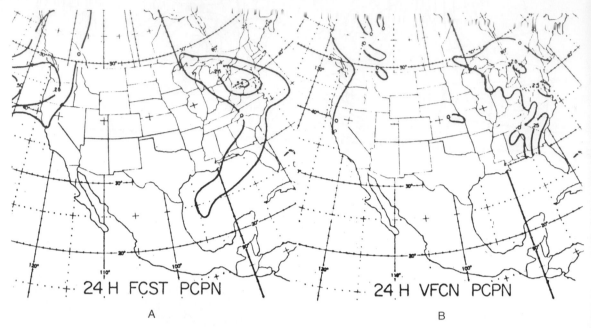

24 H FCST PCPN

A

24 H VFCN PCPN

B

**Figure 13.12**   *Forecast and observed precipitation for period 12 to 24 hours after initial time. [From Shuman and Hovermale (1968).]*

itation based on these flow patterns are improved when empirical or statistical procedures are used to supplement the NWP results. The statistical procedures consist of finding the relationship between some of the quantities that are observed or predicted numerically (for instance, the height of the 700-mb surface over a station or the difference in sea-level pressure between two stations) and the quantity to be predicted (for instance, the probability of precipitation in the next 12 hours or the maximum temperature) on the basis of past records. The quantities that are assumed to be known are called *predictors* and the quantity that is to be predicted is called the *predictand.* The problem is to find those predictors that determine with the least uncertainty the predictand and to find the relationship that enables the evaluation of the predictand for any given values of the predictors.

At the Techniques Development Laboratory of the National Weather Service in Washington the relationships for the probability of precipitation and the minimum and maximum temperature at 108 cities throughout the United States have been developed. The predictors, which were selected for physical as well as statistical reasons, include measures of the circulation pattern as given by the predicted heights of constant pressure surfaces, the humidity at various levels, and the past temperatures and precipitation at various locations. Figure 13.13 shows an example of the forecast of the probability of precipitation. This forecast was for the first 12 hours after the time of forecast. Similar charts are made and trans-

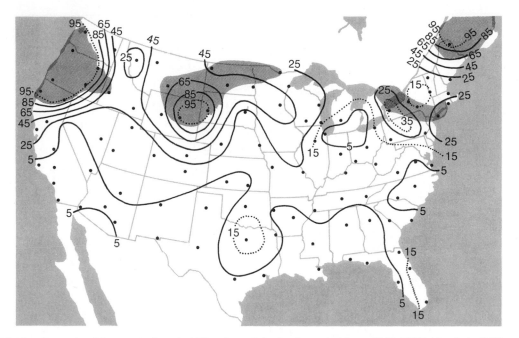

**Figure 13.13**  *Example of forecast of probability of precipitation in period from 1200 GMT, May 8, to 0000 GMT, May 9, 1970, obtained by applying statistical relationships to NWP output based on initial conditions at 1200 GMT, May 8. Shading shows area in which precipitation was observed during the forecast period. [From W. H. Klein, "Computer Prediction of Precipitation Probability in the United States," J. Appl. Meteorol. 10(5) (1971):913.]*

mitted for every 12-hour period up to 60 hours after the time of the forecast. Figure 13.14 shows the forecasts of the minimum and maximum temperatures at the 108 cities, made for periods of 24 to 60 hours in advance.

These prognostic charts, prepared at the National Meteorological Center, are transmitted to the regional and local forecast offices for their guidance. In applying them, the forecaster at each local office must modify their indications in the light of his own interpretation of the weather situation, local topographic influences, and the special needs of the public in his community. While the forecasts he issues are usually consistent with the NMC products, they contain considerably more detail, including specification of smaller-scale features, such as sea breezes or thundershowers, which the NWP filters out. The computer has not yet displaced the human from the weather forecasting process!

## 13.5  Special forecasts

In addition to the general forecasts issued for everyday use by the public, the National Weather Service makes various specialized predictions. The forecasting for air navigation has already been referred to. Frost warnings for agriculture, flood warnings, hurricane warnings, warnings of severe

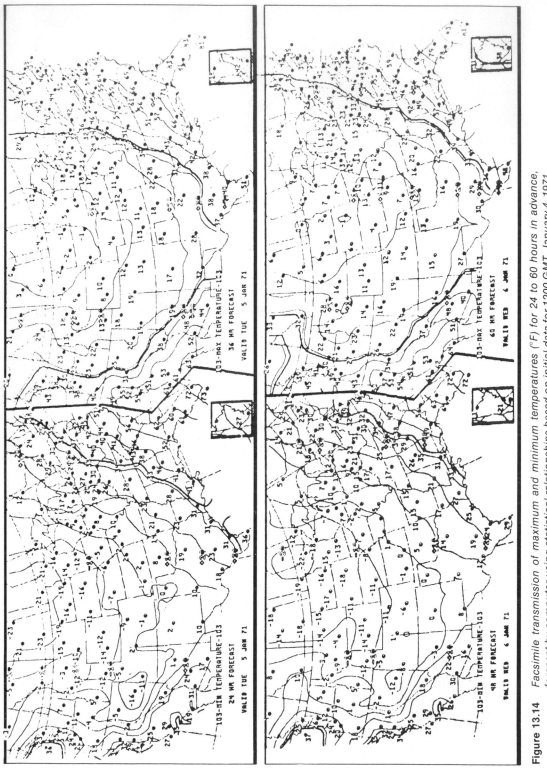

**Figure 13.14** Facsimile transmission of maximum and minimum temperatures (°F) for 24 to 60 hours in advance, forecast by computer using statistical relationships based on initial data for 1200 GMT, January 4, 1971. Inset in lower right-hand corner is northern New England. [From W. H. Klein, F. Lewis, and G. A. Hammons, "Recent Developments in Automated Max/Min Temperature Forecasting," J. Appl. Meteorol. 10(5) (1971):917.]

thunderstorms and tornadoes, fire-weather forecasts, and air pollution forecasts are other important services provided to meet the needs of special groups or the requirements of special meteorological circumstances.

These special forecast services are made by forecasters with expertise in the particular subjects, using supplemental observations. Thus, for flood forecasting, a special river and rainfall reporting network is maintained, as well as a snow survey that indicates the water content available in case of sudden thaws. The forecaster has available the past records of stream flow in response to various amounts of precipitation, snow melt, and soil and river conditions. Based on the river-stage measurements and the forecasts of precipitation and temperature, he predicts the changes in height of the river at various points along its course. For rivers with large drainage basins, the response is sufficiently slow to permit the forecast to be made with considerable accuracy well in advance. When flooding is expected, it can usually be predicted far enough in advance for protective measures to be taken—in particular, for residents of low-lying areas to move to higher ground, reducing or eliminating loss of life. Small rivers are subject to flash floods, in which the response to the heavy rain is so rapid that unless the amount of precipitation is forecast well in advance there is little time available for protective action. The Weather Service has set up a flash-flood warning procedure in a number of river basins, utilizing volunteer networks of rainfall and river observing stations that enable rapid determination of the trend of the river while heavy rain is still falling.

The hurricane warning service is headquartered at Miami, Florida. During the hurricane season for the Atlantic Ocean—June 1–November 30—a constant watch for tropical disturbances is maintained. Their early detection has been facilitated by meteorological satellite photographs. When there are indications of a developing storm, a reconnaissance aircraft, the "hurricane hunter," is sent to determine its location and intensity. As the storm approaches the coast, it comes within range of the weather radar network. The forecasting is a combination of pure extrapolation of past movement and prediction based on the influence of the large-scale circulation as forecast by NWP. Warnings of the strength of winds, the height of the storm tide, and the intensity of the precipitation (which may cause flooding) are issued for the locations that will be affected as the storm moves inland.

Forecasts of severe thunderstorms and tornadoes consist of two stages: the severe storm watch bulletin, designating the areas of probability of severe storms; and the storm warning, giving the actual position and probable movement of a damaging storm or tornado. The alert is based on the general forecast for the area. The tornado warning is based on actual visual or radar detection of the storm. For this purpose the Weather Service maintains a network of radars throughout the central and southeastern parts of the United States, where tornadoes are most frequent, and Severe Local Storm Spotter Networks of volunteer observers and communication facilities for rapid reporting and broadcasting of the sighting of storms.

When a tornado has been sighted, persons in the immediate vicinity should take cover immediately, preferably in a storm cellar or reinforced building, and stay away from windows. A basement or an interior hallway or small closet on the ground floor is best. If possible, one should curl up under a sturdy piece of furniture, which will protect against flying or falling debris.

The fire-weather service provides information about the fire danger to guide the deployment of protective forces by the Forest Service and makes wind and humidity forecasts when fires are burning, enabling the efficient use of fire-fighting men and equipment. Frost warnings enable protection of crops with orchard heaters, by flooding, or by early harvesting. In these and other applications the forecaster must take account of the topography on a very local scale and make use of his knowledge of the behavior of weather systems in relation to the specific phenomena under consideration.

## 13.6  Extended-range forecasting

The NWP tests of predictability discussed in Section 13.2 suggest that detailed day-by-day predictions are possible for two weeks at the longest. In practice, at present the accuracy of the predictions decreases from day to day, so that after three days they are poor and after five days they are little better than chance. At first sight it would seem that weather forecasts cannot be extended beyond five days at present.

However, there is valuable information we might be interested in other than the day-by-day sequence of weather. For various purposes we may find it useful to know whether next week or next month will be rainy or dry, or warmer or colder than normal for the season. The attempts at extended-range forecasting have been focused on these questions, rather than on the daily weather.

There have been many methods attempted: weather typing, statistical correlations, analogues, periodicities, and the influence of sunspots. Recently, two procedures based on physical processes have been dominant. In one, the average flow for a week or a month is studied from the standpoint of the movement of the long, slow-moving waves in the flow. This method, combined with the results of statistical studies, has been the basis for the weekly and monthly outlooks issued by the National Weather Service. The other procedure is based on the tendency for certain anomalies of the characteristics of the earth's surface—its temperature and snow cover—to produce persistent effects on the circulation of the atmosphere in subsequent months and seasons. This is particularly true when the surface anomaly, for instance, abnormally low sea-surface temperature, tends to be reinforced by its effect on the atmosphere. Some "feedback" mechanisms of this type have been found, but so far the procedure has been applied only experimentally.

The value that long-range weather forecasts would have is widely recognized. If abnormally wet or dry seasons or unusually warm or cold months could be predicted in advance, for instance, the planting and harvesting of crops, construction activities, distribution of fuels, and so forth, could be planned in a fashion that would reduce losses and increase the effectiveness of economic activities. Methods of seasonal forecasting are in a preliminary stage of development and have had slight or limited success. In economic planning, climatological records must be relied on for weather information beyond a week in advance.

## 13.7 Programs for the improvement of forecasting

Both within the weather services and in the general scientific community a large amount of research is being conducted for the purposes of improving the accuracy and extending the range of weather forecasting. Much of this effort is aimed at further increasing our understanding of the physical laws governing the structure and behavior of weather systems. For a full understanding the observations of the global atmospheric structure must be more complete and more extensive. Recognizing the tremendous scope of the task of obtaining a complete picture of the structure of the atmosphere even for one time and of understanding the complex interrelationships of the processes occurring with it at all scales, meteorologists have launched two international programs. One of them, the World Weather Watch (WWW), is a program to foster filling the gaps in the observational network and the deficiencies in the prompt communication of the observations to the data centers and from them to the forecast centers. It is being conducted by the World Meteorological Organization and the weather services of its member states. The other is the Global Atmospheric Research Program (GARP), which has been organized jointly by the International Council of Scientific Unions and the WMO. The objectives of GARP are to obtain an adequate set of observational data for the entire earth to permit the testing of NWP procedures and to develop NWP procedures that would enable the extension of the forecasts to the theoretical limit of predictability.

Implementation of the World Weather Watch has been under way since 1968. It aims at a fully coordinated system of surface, upper-air, and satellite observations. The surface observations may include automatic weather stations in remote, uninhabited areas and on buoys at sea. The upper-air network will be augmented by radiosonde observations from 100 ships while they are moving on their normal voyages. The satellite subsystem will consist of five geostationary satellites (at fixed positions above the equator) and two or three polar orbiting satellites. The geostationary satellites will give a continuous record of the development and movement of weather systems and enable determination of upper winds from measurements of cloud movements. The orbiting satellites will be equipped to make temperature soundings by means of infrared radiation measurements, in addition to providing cloud pictures.

Scientists working on the Global Atmospheric Research Program similarly envision a greatly expanded observational system. The First GARP Global Experiment (FGGE) will attempt for the first time to obtain a complete specification of the state of the atmosphere over the entire earth for a whole year. The FGGE will have the following four objectives:

1 To obtain the understanding of atmospheric motions required for the development of models for extended-range forecasting, general circulation studies, and climatic prediction
2 To assess the ultimate limit of predictability
3 To develop improved methods of accumulating and summarizing meterological observations, including nonsynchronous data, to specify the initial condition for NWP
4 To design the optimum observing system for routine NWP

The FGGE is expected to start in 1976. In addition to the basic WWW system, the FGGE will make use of experimental satellites to determine the state of the ground surface and the location of precipitation areas, a fleet of about 500 constant-level balloons to monitor the upper-air temperature, pressure, and winds in the Southern Hemisphere, a network of buoys to provide data over the oceans, a fleet of 200 carrier balloons flying at high levels (30 mb) and releasing dropsondes (radiosondes carried downward on parachutes) to give winds and upper-air temperatures over tropical regions, and additional ocean ship stations.

Preliminary studies have already been carried out or are under way to determine the characteristics required for the FGGE observational system. In particular, the relationship between subgrid-scale phenomena, such as cumulus convection, and large-scale circulations is being investigated. For this purpose a preliminary large-scale experiment, the GARP Atlantic Tropical Experiment, will be conducted in 1974. The transfer of energy and water vapor from ocean to atmosphere, the release of latent heat during the formation of clouds and precipitation, and the transfer of this energy to the larger flow patterns will be studied.

Through programs like WWW and GARP, the potential for benefit to society from the understanding of our atmospheric environment will be realized.

## Questions, Problems, and Projects for Chapter 13

1　Discuss the reasons why a prediction of tomorrow's weather cannot be carried out with the same accuracy as the prediction of eclipses of the sun and moon years ahead.

2　In most localities the observed maximum temperature, minimum temperature, and precipitation for each day and the forecast of these quantities for the following day are published in the daily newspaper or broadcast on television and radio.

　　a　Keep a record of the observed values and the forecasts for a month.

　　b　As measures of the accuracy of the temperature forecasts, evaluate the average difference between the observed and forecast values, separately for the maximum temperature and the minimum temperature.

　　c　As a measure of the accuracy of the precipitation forecasts, count the number of days for which the precipitation forecast was right. This should comprise the days precipitation was forecast on which it did occur and the days precipitation was not forecast and none occurred. Express this as a percentage of the total number of forecasts.

# FOURTEEN

## Air
## Pollution

### 14.1    Types and sources of air pollution

Almost all human activities introduce contaminants into the atmosphere. Natural processes also introduce other substances than those we regard as the constituents of pure air. Besides nitrogen, oxygen, argon, carbon dioxide, water in its various phases, and the trace gases that form a permanent part of the atmosphere, the atmosphere always contains emanations from growing or decaying vegetation, salt from sea spray, dust from blowing soil and sand storms, smoke from lightning-caused fires, and gases and fumes from volcanic eruptions. Except locally or temporarily, these natural substances are in low concentrations and have no harmful consequences. The artificial injection of contaminants by human activities, however, has caused illness and death to humans, impaired agricultural productivity, and damaged property. The levels that pollution has reached have already aroused worldwide concern, and the possibility that irreversible consequences will occur unless the emissions of pollutants are reduced or eliminated has been suggested.

Air pollution may be defined as the presence in the atmosphere of substances that are toxic, irritant, or otherwise harmful to man or damaging to vegetation, animals, or property. The combustion of fuels to produce energy is the principal source of air pollution. Until recently the problems of air pollution were concerned principally with the products of coal

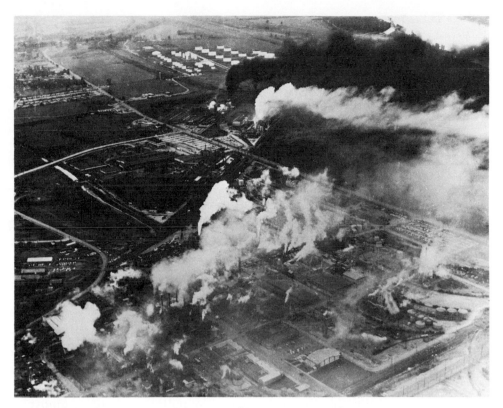

**Figure 14.1** *Air pollution from an industrial complex. [Courtesy of NOAA.]*

combustion—namely, soot, sulfur dioxide, and fly ash—and the effluents from such industrial operations as smelters and steel mills. At high humidities water drops condense on coal smoke to form the very dense "pea soup" fogs that were characteristic of such cities as London, Pittsburgh, and St. Louis before they enacted legislation requiring the use of clean fuels and efficient means of combustion. The word *smog*, formed by combining the words *smoke* and *fog*, was originally used to designate this type of air pollution. It now has come to be used as a synonym for general air pollution whether or not it involves smoke or occurs at high humidity.

In recent years petroleum derivatives have replaced coal as the principal source of energy in many places: fuel oil and natural gas for heating and electric power generation; and gasoline, kerosene (for jet airplanes), and diesel oil for transportation. The use of these fuels has led to a new type of air pollution, in which photochemical reactions play an important role. A photochemical reaction is one in which the absorption of radiation initiates or facilitates chemical changes. In photochemical smog, nitrogen dioxide absorbs solar radiation to initiate a chain of reactions. In the presence of hydrocarbons the process results in the formation of ozone and other highly reactive, irritating, and toxic substances. Because it

**Table 14.1** Total Emissions of Primary Gaseous Pollutants
(millions of tons per year)

| Pollutant | U.S. | World |
|---|---|---|
| Sulfur dioxide ($SO_2$) | 29 | 146 |
| Carbon monoxide (CO) | 102 | 270 |
| Hydrocarbons ("HC") | 32 | 88 |
| Oxides of nitrogen ("$NO_x$") | 21 | 53 |
| Total | 184 | 557 |

**Table 14.2** Major Sources of Air Pollution in the United States (millions of tons per year)

| Source | $SO_2$ | CO | HC | $NO_x$ | Particles | Total |
|---|---|---|---|---|---|---|
| Transportation | 1 | 93 | 21 | 10 | 2 | 127 |
| Industry | 10 | 3 | 7 | 4 | 10 | 34 |
| Electric power plants | 14 | 1 | <1 | 5 | 5 | 25 |
| Domestic heating | 4 | 3 | 2 | 2 | 2 | 13 |
| Refuse disposal | <1 | 2 | 2 | <1 | 2 | 6 |
| Total | 29 | 102 | 32 | 21 | 21 | 205 |

was first observed in Los Angeles, photochemical smog is frequently referred to as *Los Angeles–type smog.* The coal-smoke smog is called *London-
type smog.*

Estimating the total emissions of pollutants into the atmosphere involves
many uncertainties. There is no strict census of the amounts emitted even
in industrialized nations, such as the United States. Rough estimates of
emissions of gaseous pollutants for the United States and the world are
presented in Table 14.1 Only the total annual values are given, as the
seasonal variation of the figures on which they are based, such as the
consumption of coal, are not available. In Table 14.2 the amounts emitted
by various types of sources are shown for the United States. It is seen that
the main contributor of sulfur dioxide and particulates in the atmosphere
is industry, including electric power generation, whereas carbon monoxide
and hydrocarbons come chiefly from vehicles used in transportation.

After the pollutants have been emitted, they spread by being carried
away by the winds and being mixed horizontally and vertically by turbulent diffusion and convection. Ultimately, they are removed from the
atmosphere by processes we shall discuss in a later section. The average
length of time that pollutants remain in the atmosphere varies. For large
particles and for reactive gases it may be only minutes or at most hours.
For sulfur dioxide it is estimated to be a few days, but for carbon monoxide
it may be several months. If the pollutants were emitted uniformly over
the earth's surface and mixed throughout the atmosphere, even those with
relatively long lifetimes would have concentrations so low as to be incon-

sequential at present average emission rates. It is because the emissions occur in small areas and it takes time for them to spread that objectionable concentrations accumulate.

## 14.2 Factors affecting the concentration of air pollution

The primary determinant of air-pollution concentration is, of course, the amount of contaminants emitted into the air. But our experience tells us that even though the same sources pour out pollution day after day, sometimes the air is relatively clean and sometimes it is very polluted. The concentration of pollution depends on the weather conditions. In addition, for the same amount of emission and the same meteorological conditions, the air-pollution concentration is influenced by the geometric configuration of the sources, including the height above ground of the emission and the area over which the sources are distributed. If the emission is at a considerable height, the pollutants become diluted by the time they are transported to the ground by eddy diffusion. If the emission is spread out over a large area, the downwind concentration will be smaller than if it all comes from a small area, equivalent to a point source. The factors that affect the concentration of air pollution are thus (1) the total amount of pollution emitted; (2) the meteorological conditions; and (3) the configuration of the sources. As an aside we may note that for pollution control we can alter (1) and (3). Although weather modification has been proposed as a means of reducing air pollution, all such proposals made up to the present time have been shown to be impractical.

The amount of pollution emitted in industrial, commercial, and domestic activities and the problem of reducing it are essentially subjects for engineering studies. However, meteorology enters into these questions insofar as the estimation of the capacity of the air to dilute the pollutants determines the allowable emissions to which the engineers should limit the equipment they design. Similarly, the height of smokestacks, the shape of factories, and the layout of commercial areas, housing developments, and cities are the provinces of architecture, engineering, and urban planning, but meteorological factors should be taken into account if the pollution coming from these structures is to be minimized.

The various types of sources are for convenience treated as point sources, line sources, and area sources. The smokestack of an electric power generating plant is an example of a point source. The exhaust pipe of an automobile similarly constitutes a point source; however, because motor vehicles move in rapid succession along them, streets and freeways act essentially as line sources. The chimneys of houses and apartment buildings, regarded individually, are also point sources, but they are so close together that they merge into a continuous area source. The emissions from a point source are expressed in terms of mass per unit time, for instance, kg s⁻¹; from line sources as mass per unit length per unit time, for example, kg m⁻¹ s⁻¹; and from area sources mass per unit area per unit time, kg m⁻² s⁻¹.

The meteorological factors that affect the pollution concentration are (1) the stability of the air as determined by the vertical variation of temperature and (2) the direction and strength of the wind. These factors determine how fast the contaminant is diluted by mixing with environmental air after it leaves the source. The hydrostatic stability controls the rate at which up-and-down motions mix the pollutants with clear air from above. The wind speed determines into how much air the contaminants are initially mixed, and the irregularities of the wind speed and direction govern the rate at which the pollutant spreads horizontally as it is carried downwind.

The effect of stability has already been discussed in Chapter 4. If there is a superadiabatic lapse rate, vertical mixing occurs readily and pollutants are diluted rapidly. If there is an inversion, the great stability completely suppresses vertical mixing and dilution can occur only to the extent that there is horizontal admixture of clean air.

The effect of stability is rendered strikingly visible by the behavior of smoke issuing from a stack. Figure 14.2 shows schematically the various types of smoke plumes associated with various types of variation of temperature with height. In diagram A the smoke is emitted into an inversion layer. The stability prevents diffusion up and down, so that the only spreading of the smoke is sideways. Since the plume is thin in the vertical and is V-shaped in the horizontal, it has been likened to a lady's fan; hence the name "fanning." In diagram B there is an adiabatic lapse rate in the lower layer, topped by an inversion. Downward mixing goes on readily, but the inversion limits the upward mixing. This configuration arises when there is an inversion at the ground at sunrise and the heating from solar radiation produces an adiabatic or slightly superadiabatic lapse rate through a progressively thicker layer until the level at which the smoke plume has been confined until then is reached. After this level is reached, the pollution mixes downward rapidly, "fumigating" the ground, which has until this time been protected from the pollution by the inversion. This process may result in an abrupt increase of the pollution concentration at the ground to a high value.

When there is a superadiabatic lapse rate through a deep layer (diagram C), the smoke is carried up and down by convective currents, forming a *looping* pattern, and is rapidly diluted by the intense vertical mixing. With a deep adiabatic layer (diagram D), vertical mixing also goes on freely, but the turbulent motions that are induced by irregularities of the ground and shearing of the wind are not amplified by instability. The vertical spreading and lateral spreading are about equal, and the smoke plume

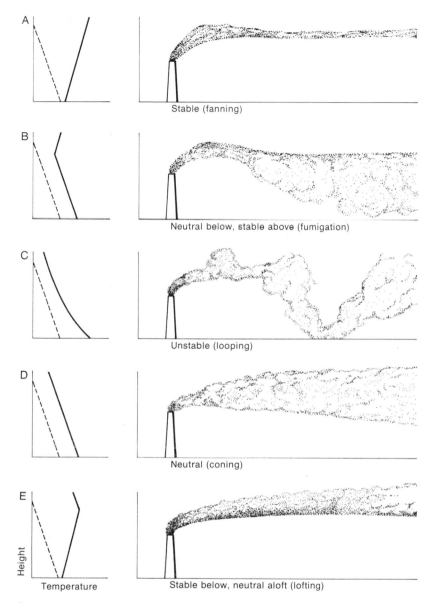

**Figure 14.2** *Smoke-plume patterns characteristic of various types of variation of temperature with height. The dashed curves in the diagrams on the left represent the adiabatic lapse rate.* [*After D. H. Slade, ed.,* Meteorology and Atomic Energy 1968 *(Oak Ridge, Tennessee: U.S. Atomic Energy Commission, 1968).*]

resembles a cone. In the remaining possible configuration, diagram E, the smoke is emitted at the top of an inversion layer, where it is kept from mixing downward but spreads upward. This tendency to be carried aloft but not to the ground has been termed *lofting*.

The vertical diffusion of emissions from line and area sources is influenced in the same way by stability. With the increased horizontal dimen-

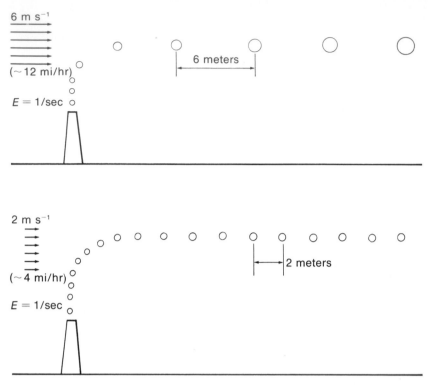

**Figure 14.3**  *Effect of wind speed on pollution concentration. The size of the "bubbles" is intended to show the amount of clean air with which the pollutants in them is mixed. The stronger wind in the upper diagram leads to lower concentrations both because the bubbles of pollution are spread farther apart and because they are diluted more by mixing.*

sions of these sources the dilution due to horizontal mixing is reduced. Over an area source, for instance, lateral stirring mixes the pollution from one part of the source with that from other parts of it, rather than with clean, unpolluted air. The presence of an inversion over a large area source leads to very high concentrations.

The way in which the wind speed affects the concentration of pollutants is illustrated in Figure 14.3. The pollution is assumed to be entering the atmosphere at the rate of one unit per second, each unit being represented for convenience by an individual bubble. In the upper diagram the wind is 6 m s$^{-1}$, and the bubbles are therefore spread six meters apart by the wind, whereas the wind in the lower diagram is one-third as strong and the bubbles are one-third as far apart. Due to the direct effect of wind speed, the concentration is three times as great with the 2 m s$^{-1}$ wind as with the 6 m s$^{-1}$ wind.

A second effect of wind speed, turbulent mixing, is represented by the size of the bubbles as they move downstream. The stronger wind is more turbulent, so that the polluted air mixes more rapidly with the air around it and become more dilute; with the light wind there is less turbulence and the concentration remains high. Turbulence manifests itself in fluctuations of wind direction as well as speed. The variations in speed produce mixing along the average wind direction, and the variations in direction lead to side-to-side mixing and to up-and-down mixing. As has been discussed, the amount of vertical mixing is largely controlled by the hydrostatic stability. However, for the same stability, vertical mixing varies with the wind. In light winds the mixing in all directions is much less than in strong winds.

From the foregoing discussion we see that for strong winds and large lapse rates the diffusion of pollutants is rapid. Under these conditions high concentrations of pollution will not occur except very near intense sources. When the wind is very light and an inversion is present, however, the pollution diffuses slowly and high concentrations develop if there are sources of sufficient magnitude.

Light winds and inversions occur together occasionally everywhere in the world, but there are some places where they are much more frequent than others. The frequency with which they occur, and in particular the length of time they persist, are measures of the *air-pollution potential* of a place.

## 14.3 Evaluation of air-pollution potential

Although the frequency with which the combination of light winds and inversions occurs is different for various parts of the world, there is no place where it does not occur. Even the windiest regions on earth would occasionally have high concentrations of pollution if sources were present. The climate of a place determines how much of the time there will be severe smog if sources are present. The intensity of the sources determines whether it will ever be severe. Unless the pollution sources are sufficient, even the most adverse meteorological conditions will not lead to adverse concentrations. The severity of smog depends on both the sources and the meteorological conditions.

Persistent light winds and inversions occur together principally during periods when a stagnant or slow moving anticyclone is present. In temperate and high latitudes the fall and winter are the seasons when such periods are likely. Winds are light near the center of an anticyclone, and the horizontal outflow from the high-pressure center produces sinking motion and adiabatic heating aloft. At the same time the temperature of the ground and the air immediately above it falls because of radiation into clear skies. This cooling below and heating above produces an inversion so strong that it is not destroyed by the heating of solar radiation during the short days of autumn and winter.

Similarly, the semipermanent subtropical high-pressure areas (see Sections 7.7 and 12.1) have light winds and inversions associated with them. The subsidence is concentrated on the eastern sides of the anticyclones so that the continental areas affected by the inversion produced by it are the west coasts, at latitudes ranging from 40° or more in both hemispheres almost to the equator. In winter, invasions of storms, fronts, and cold, unstable air masses interrupt the anticyclonic influence, and periods of high smog potential are less frequent and shorter in duration. But in summer the trade-wind inversion and light winds are present continuously at subtropical west coasts. From the standpoint of pollution potential such places as Casablanca, Morocco, Capetown, South Africa, Santiago, Chile, and Los Angeles, California, are among the worst possible locations for cities.

The pollution potential of a city can be estimated in terms of its size, the wind speed, and a quantity called the mixing height, representing the height through which the pollutants will be completely mixed in crossing the city. If the pollution is thoroughly mixed up to $H$, the concentration $C$ of pollution at a distance $L$ downwind from the outskirts of the city, according to a simple "box" model, is given by

$$C = QL/UH \qquad (14.1)$$

where $Q$ is the emission rate per unit area, assumed to be uniform, and $U$ is the average wind speed. The formula agrees with our expectations that the concentration is higher the more intense and more extensive the sources over which the air has passed, the slower the movement of the air across the sources, and the smaller the depth of air into which the pollution mixes. In this simple model it is assumed that lateral mixing does not produce a decrease in concentration and that there is no removal of pollutants. Although the model is a gross oversimplification, it focuses attention on the factors that would enter a more accurate expression for computing the concentration of pollution, namely, the properties of the pollution sources, $Q$ and $L$, and the meteorological factors, $U$ and $H$.

In the daytime the mixing height is determined by the depth of the layer through which the sun's radiation has established an adiabatic lapse rate. On clear nights the radiational cooling might be expected to establish an inversion and reduce $H$ almost to zero. However, it has been found that over cities the heat retained by the buildings or added by human activities maintains a lapse of temperature through the lowest 100–200 m. Thus even at night there is some dilution of pollutants over cities because of vertical mixing. The values of $H$ range from 100 or 200 m at night and on some winter days to 3 km or more on summer afternoons.

The wind in most cities undergoes a similar diurnal and seasonal variation. Usually, it is lightest at night and in the early morning and has a maximum in the afternoon, when vertical mixing transfers momentum downward from higher levels where the wind is stronger because it is not

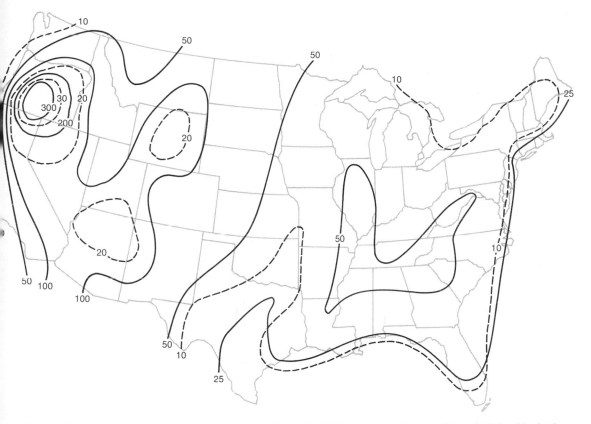

**Figure 14.4**  *Values of median annual concentration ratio C/Q in the morning for cities of 10-km (dashed lines) and 100-km (continuous lines) along-wind dimensions. [After G. C. Holzworth, Paper presented at the Second International Conference on Air Pollution, Washington, D.C., December, 1970.]*

impeded by surface friction. In regions such as sea coasts and valleys the diurnal wind regime will be dominated by topographic influences, especially when the general flow is light. Usually, these influences have the same pattern—light winds at night and maximum winds in the afternoon.

Most primary pollutants have their diurnal maximum concentration an hour or two after sunrise, when human activities have increased $Q$ to a high value but $U$ and $H$ are still small. As indications of the daily pollution potential, the annual median values of the concentration ratio $C/Q$ for the United States (except Alaska and Hawaii) are given in Figure 14.4 for two values of $L$. The values in this figure were obtained by using a somewhat more sophisticated model than that represented by equation (14.1); for this reason the values for the 100-km city are not exactly ten times those for the 10-km city.

Of more interest than the average values is the frequency of extremely high values of concentration. Figure 14.5 shows the values of $C/Q$ that would have been exceeded on 10 percent of the mornings each year. Here

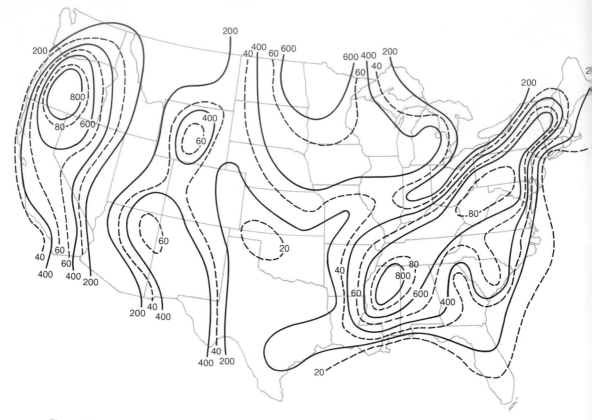

**Figure 14.5** *Values of concentration ratio C/Q exceeded on 10 percent of the mornings annually, for cities of 10-km (dashed lines) and 100-km (continuous lines) along-wind dimensions. [After Holzworth (1968).]*

we see that whereas a 100-km city would experience values of $C/Q$ greater than 100 one-half of the time only over the western quarter of the country, it would have values in excess of 400 during 10 percent of the time in much of the East as well as the West.

It appears from these maps that the area of the worst pollution potential is southern Oregon, where the smog in a 10-km city would be almost as bad one-half of the time as that which Los Angeles experiences on the worst 10 percent of the days. In turn, Los Angeles has more than twice the pollution potential of New York.

It should be noted that these charts give morning values. Only if the potential remained high long enough for the air to move across the city would the actual pollution concentration reach these values; and in any case if in the afternoon the wind is strong enough and the layer of turbulent mixing is deep enough the concentrations will rapidly decrease. The effects on health are usually dependent on the dosage, that is, on the concentration

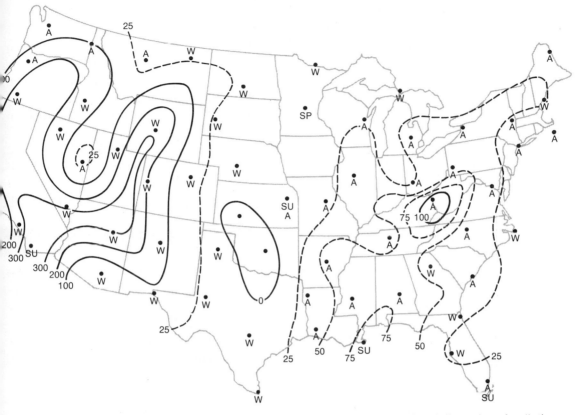

**Figure 14.6**    *Number of days in five years for which conditions favored reduced dispersion of pollution continuously for at least 48 hours. Season in which most such days occurred at each station is shown by W (winter), SU (summer), SP (spring), or A (autumn)* [*Holzworth (1968).*]

multiplied by the duration that the concentration has the particular value. If the concentration has a very high value but is sustained only for a short time, its effects will not be as bad as those of concentrations having a lower peak value that is sustained for a long period. Figure 14.6 presents an estimate of the frequency of sustained high concentrations. It gives the number of "episode days" in a five-year period, where an episode is defined as a period of at least two days in which the mixing height and wind speed were sufficiently low to lead to high pollution-concentration ratios. The area of the highest number of episode days is in the West, from the Rocky Mountains westward. San Diego has the largest number. Los Angeles, which presently has much more pollution because of its larger size (greater $Q$ and $L$), has less than one-half as many, primarily because its afternoon winds tend to be a little stronger. It is clear that in many parts of the western United States if cities grow to the size of Los Angeles they also will be subject to severe smog problems.

## 14.4 Effects of air pollution

At the beginning of this chapter some of the adverse effects of air pollution were mentioned. It would seem almost unnecessary to go into details about these effects in order to justify the imposition of measures to reduce it. However, the reduction of pollution can be achieved only at considerable expense, and the justification of these expenditures requires an analysis of the benefits to be achieved.

The effects of greatest concern are those on human health. There have been several instances of large numbers of deaths attributed to air pollution. The largest number occurred in the London disaster of December 5–9, 1952. During this period an anticyclone stagnated over England, producing a strong inversion and nearly calm conditions. The temperature was lower than normal, leading to augmented emissions from the millions of domestic fires and industrial chimneys. A persistent dense smog in the true sense of the word developed, with the coal smoke providing hygroscopic nuclei on which the condensation took place. Measured concentrations of particulates and sulfur dioxide were many times the values usually found in the polluted London air. Before the smog was cleared out by the approach of a frontal system, a great many people became ill, and there were over four thousand more deaths than the number normally taking place during a five-day period at that time of year.

The public reaction to this terrible smog "incident" led to the passage by Parliament of the British Clean Air Act of 1956, which required the use of "smokeless" fuels and improved combustion methods. Similar procedures had previously led to a reduction of smokiness of some American cities, notably, Pittsburgh and St. Louis; and since the adoption of this law, the visibility in London has improved remarkably. No longer does London experience the "pea soup" fogs that used to typify it.

In terms of mortality, the first important occurrence of smog in the United States was the Donora disaster of October 26–29, 1948. Donora, Pennsylvania, is a small industrial town in the valley of the Monongahela River, about 20 miles southeast of Pittsburgh. A persistent inversion developed on October 26 because of the stagnation of an anticyclone. Into the calm, foggy air the zinc plant, steel mill, and other industrial establishments poured their effluents for three days, producing increasing concentrations of sulfur dioxide and other contaminants. By the third day almost one-half of the 14,000 people in the town became ill and twenty died.

There had been many occurrences of high pollution concentrations before the disastrous ones in London and Donora, and there have been many since. In some of them it has been established that deaths due to the smog occurred, but in most cases it is uncertain whether any increase in mortality could be ascribed to the pollution. Nevertheless, the few instances in which pollution has been shown definitely to have caused death, together with evidence that some of the contaminants in the atmosphere are toxic, clearly show that air pollution is a hazard to human health.

Many substances commonly found in polluted air are known to be responsible for illness when they are present in high enough concentrations. Usually, the concentrations at which they have harmful effects in laboratory tests are much higher than those observed in the atmosphere. Sulfur dioxide, carbon monoxide, hydrocarbons, nitrogen dioxide, lead, sulfate particles, and ozone are among the pollutants that are known to have adverse effects on humans.

The effects of pollutants on vegetation have long been recognized. Sulfur dioxide has received the most attention, but in recent years the products of photochemical smog in very low concentrations have been shown to cause injury to plants. Photochemical smog damage has been found in the vicinity of most of the major metropolitan areas of the United States. In the San Bernardino Mountains of California, the pine and fir trees of the national forest are dying because of smog from the Los Angeles area. The damage to vegetation includes the burning or marring of leaves, the stunting of growth, a decrease in the size and yield of fruits, and the destruction of flowers.

The total annual cost of damage to vegetation by air pollution, including the losses to farms, forests, and domestic gardens, has been estimated to be almost $1 billion.

The effects of pollution on materials and structures also have been known for a long time. Corrosion of metals, soiling and erosion of building surfaces, discoloration of paints, soiling and weakening of textiles, deterioration of sculpture, paintings, and other works of art, and weakening of rubber goods, including automobile tires and hoses—all are caused or accelerated by air pollution. The total cost of cleaning, protecting, and replacing materials damaged by smog is not known, but estimates indicate that it amounts to several billion dollars a year in the United States.

One of the most conspicuous manifestations of air pollution is the reduction of visibility. This reduction is due to the scattering of light by particles. It is an effect of particulate pollution only and not of gaseous contaminants. Thus, the striking improvement in visibility that has taken place in Pittsburgh, St. Louis, and London since the adoption of measures to reduce the amount of coal smoke must not be interpreted as a complete solution to their air-pollution problems. The gaseous contaminants were not necessarily reduced thereby, and they may cause more serious effects than the particulates.

The reduction of visibility by pollution may be sufficient at times to interfere with transportation. Airplanes may have to operate on instrument flight rules instead of visual flight rules, with consequent costly slowdown of landing operations.

Pollution also reduces the amount of solar radiation reaching the ground.

This reduction, particularly in the ultraviolet, may contribute to the adverse effects of pollution on human health.

Other effects of pollution on the weather will be discussed in the next chapter, in which we shall consider both the intentional and the unintentional influences of man on the atmosphere.

## 14.5 Control of air pollution

The obvious way to control air pollution is to keep the contaminants from entering the atmosphere. There is no way to do this completely. All human activities produce wastes, some of which automatically enter the air. The production and use of energy, the refining of minerals, the various modes of transportation, and such commonplace actions as house-painting and dry cleaning—all use the air to carry away waste products. Since complete elimination of these activities is unreasonable, we must seek ways to carry them out that lead to acceptably low concentrations of pollutants.

Because the objective of commerce and industry is to yield a profit for owners or stockholders, no individual company will purchase expensive equipment to reduce its emissions of pollutants unless its competitors are required to do likewise. Thus, control of pollution cannot be expected on a voluntary basis. It must be imposed by law, so that all businesses are required to meet the same standards. Furthermore, since pollution is carried by the wind, which does not respect municipal boundaries, and since businesses in one city or state compete with those in others, the laws controlling air pollution must be at a sufficiently high level of government. Measures passed and enforced by state and national governments have a better chance of producing the desired air quality than those imposed by municipal or county agencies. On the other hand, because situations may be more acute in some cities than in others or in rural areas, there must be provision for stricter standards in some localities than apply to the general region.

Laws may specify the allowable limits of concentrations of pollution in the air or the allowable rates of emission into it. More often, the legislation does not set up specific limits, but instead gives an enforcement agency the power to do so after determining on the basis of expert investigations and hearings what the limits should be and whether their attainment is feasible. A very successful procedure has been the permit system, in which installation and operation of equipment that produces pollution is allowed only after it has been demonstrated that the emissions from the equipment will meet the standards that have been set. It has been the policy of governmental institutions not to establish requirements that cannot be met with existing technology at reasonable expense. This policy has been criticized by some who consider that the only way to stimulate the development of ways to reduce pollution adequately is to base the rules on the levels needed for acceptable air purity, independent of whether we have the means at present to attain these levels.

In the early years of pollution control, measures were left almost exclusively to municipalities and were concerned principally with establishing tolerable levels of pollution in the vicinities of large industrial plants. These levels were usually expressed in terms of the maximum allowable concentration at the ground of such contaminants as sulfur dioxide. To meet these requirements factories and electric generating plants built higher and higher smokestacks as the amount of their emissions increased. Another criterion that was used was the opacity of the effluent as it left the stack. By limiting the darkness of the smoke, the amount of particulate matter entering the atmosphere was controlled to some extent.

As the problem became more general, municipalities and counties have joined to form air pollution control districts, and the control measures have attempted to limit the emissions from all sources. With the spread of pollution from one area to another and with the recognition that such ubiquitous sources as the automobile contribute a large proportion of community air pollution, state governments and the federal government have launched air pollution control programs. At present virtually every one of the fifty states in the United States has pollution control agencies. The federal government, which began with a feeble program emphasizing the exchange of information under the Clean Air Act of 1955, has undertaken vigorous control and enforcement activity under the Air Quality Act of 1967 and the Clean Air Amendments of 1970. Under these laws the Environmental Protection Agency (EPA) is empowered to establish air quality control regions, set ambient air quality standards, require local and state authorities to devise plans to bring the air quality in their regions up to the standards, and, in those instances in which the local and state programs are inadequate, to inaugurate measures of its own to enable the achievement of the standards.

With respect to motor vehicles, EPA has set emission standards for 1975 at approximately 10 percent of the emissions of carbon monoxide, hydrocarbons, and oxides of nitrogen permitted for the 1970 cars. If these standards are met, the amounts listed for transportation in Table 14.2 will undergo a marked decrease as new automobiles replace old ones on the road, even with a continued increase in the number of cars.

The continuing rise in population constitutes a threat that pollution will continue to increase in spite of the control measures. Not only might the growing demands for materials and energy offset the reduction in per capita contribution to air pollution because of control, but the need to use less-rich ores and less-pure fuels may make it impossible to meet the demands and the emission limits at the same time. The energy requirements may be met without turning to dirtier fossil fuels by using nuclear power, which is almost completely free of atmospheric pollution except for the possibility of accidental release of radioactive material. Very thorough measures to guard against accidental release are incorporated in the design of nuclear reactors, and the chances of its happening are considered to be extremely small, much smaller than the chances of plane crashes or railroad accidents. In routine operation, nuclear reactors emit into the air a negligible amount of radioactivity. The amount is so small that someone at the boundary of a reactor site would experience an increase of less than 1 percent of the normal background radioactivity. There are other considerations that need to be evaluated in connection with the decision of whether or not to replace fossil fuel by nuclear reactors as an energy source, such as the problem of disposing of the highly radioactive spent fuel. From the standpoint of air pollution, the shift would be desirable.

The reduction of emissions and the restoring of tolerable air quality appear technically feasible if we are willing to undertake the costs of doing so. These costs may in the long run include, in addition to higher prices for commodities to cover the expense of their production in a pollution-free manner, the acceptance of a way of life that does not depend as much on energy-consuming appliances and that provides for a stable population level.

It is frequently suggested that it may be easier or less costly to change the meteorological conditions that limit the capacity of the air for pollutants. Various proposals have been made, including large fans to augment the air motion when the wind is too light, "punching holes" in inversions when they are present, using the waste heat of electric generating plants to augment convective mixing, painting the roofs of buildings in alternate city blocks black and white so that the differences in solar heating produce convection, and cutting down the Rocky Mountains to permit more rapid flow of air across the United States. As a rule these proposals, except for those depending on heat from the sun or the waste heat from power plants, would require such tremendous amounts of energy that its production, in addition to being very costly, would introduce more pollution into the air than the procedure could remove. The proposals for the use of solar energy or waste energy usually do not take sufficiently into account the magnitude of the problem. Computations have shown that the modification of the temperature structure or air flow to remove pollution by any of the methods so far proposed is impractical. Until a practical procedure for doing so is proposed, the only way to control pollution remains the reduction of emissions at the source.

### Questions, Problems, and Projects for Chapter 14

1 Make a "census" of the principal sources of air pollution in your community, listing each major type of source and each kind of contaminant that comes from it. What steps could be taken to reduce the emissions from each source?

2   The number of people who become ill or die in a city each day depends on several factors, in addition to the amount of air pollution present. How would you separate out the effects of the other factors in order to tell whether air pollution is responsible for any increase in illness and death?

3   Compare the factors that influence lateral (sideways) mixing with those controlling vertical mixing, and discuss the effectiveness of the two types of mixing in reducing the concentration of pollutants from point sources, line sources, and area sources.

4   Explain why Los Angeles, California, was the first place to have high concentrations of photochemical smog. Why is it less frequent at cities in higher latitudes?

5   Discuss the meteorological factors that should be considered in establishing an air quality control region.

6   How would you interpret the proposal to "punch holes" in an inversion? What would be required in order to eliminate an inversion over an area?

# FIFTEEN

## Man's Influence on the Atmosphere. Inadvertent and Intentional Weather Modification

### 15.1 Effects of human activities. The weather in cities

The activities of man alter atmospheric conditions in three ways: (1) by changing the character of the earth's surface; (2) by adding energy to it from artificial sources; and (3) by adding matter to it. The modification of the surface affects the way in which solar radiation is absorbed and retransmitted to the atmosphere and changes the frictional resistance to the wind. The combustion of fossil fuels (coal, petroleum derivatives, and natural gas) heats the air and adds particulate and gaseous contaminants. Nuclear reactors likewise add heat and contaminants to the air. Many other agricultural, industrial, commercial, and domestic activities introduce matter (pollutants) into the atmosphere.

In early times when man subsisted by hunting other animals and collecting seeds and berries the effects were minimal, being confined to the immediate vicinity of the fires he maintained for warmth and cooking. With the beginning of agriculture, 8,000 to 10,000 years ago, the effects started to become more extensive both because of the substitution of cultivated crops for natural fields and forests and because of the rapid increase in human population, which it made possible. The effects of agriculture presumably were felt principally in the immediate vicinity of the modified terrain, but since forests were chopped down over tremendous areas the changes that occurred in the exchange of heat, moisture, and momentum between the ground and the atmosphere were on a very large scale. As these transformations took place mostly before the beginning of systematic observations, the amount of change in the resulting weather is not known.

Following the spread of agriculture, trade and commerce led to the development of larger and larger towns and cities, which had their own characteristic effects on the weather. The industrial revolution accelerated the concentration of people in cities, and the increased use of fossil fuels (coal, petroleum derivatives, and natural gas) aggravated the influence of cities on the atmosphere. Since most of the growth of cities has taken place since the beginning of weather records and since the effect of cities on the weather can be estimated by comparison with surrounding rural areas, the extent of their effect is fairly well known.

The most definite influence of cities is on the temperature. Cities are on the average 1°C to 2°C warmer than surrounding rural areas both day and night and in all seasons of the year. Several factors contribute to this phenomenon, with different ones dominating at different times. The main influence during the day in the growing season is the relative absence in cities of evaporation and transpiration of water from soil and plants. These processes use a large part of the insolation in rural areas, and in their absence the greater amount of solar energy raises the city's temperature. At night the city is kept warmer than its surroundings by the reemission of heat that was absorbed by streets and buildings because of the greater conductivity and heat capacity of concrete, asphalt, and brick than of soil and vegetation. The energy used for industrial processes, heating, cooling, and illumination of buildings, other industrial and domestic purposes, and transportation ultimately is added to the outside air and raises its temperature. The energy of human metabolism also contributes a little. The dark and irregular surfaces of buildings, with vertical walls that permit multiple reflections of incoming sunlight, reduce the amount of solar radiation reflected back toward space, thereby increasing the energy available to heat the air over cities.

The area affected by the warmth of a city is referred to as an "urban heat island." Because the air moving from rural areas to cities on clear nights is heated, surface inversions usually do not occur over the urban heat island. Instead, there is a normal lapse rate through the lowest one hundred meters or so, above which the usual nocturnal inversion occurs.

The winds in cities are generally lighter than in the country because the buildings of cities serve as obstacles to the wind, increasing the frictional resistance. On the other hand, there is a channelling effect when strong winds blow along streets with tall buildings on either side, and the winds may be augmented in this circumstance. The irregularity of the surface causes more turbulent fluctuations of the wind over cities.

The effect of cities on precipitation is small, and because of the great variability of rainfall it is hard to determine. Analysis of the precipitation records at a number of cities throughout the world has shown that wherever a statistically significant difference exists there is about 10 percent more precipitation in and downwind of a city than in its surroundings, particularly in the direction from which the prevailing wind blows. The reason for this increase due to the presence of the city has not been determined. It is thought that it is more likely caused by increased vertical motion due to the urban heat island or the convergence produced by frictional resistance than by a larger number of condensation or ice nuclei introduced by pollution sources.

Pollution sources, however, are definitely responsible for a smaller amount of solar radiation reaching the ground in cities than in surrounding rural areas, a reduction in the visibility (visual range), and a greater frequency of fog. The insolation is reduced as much as 20 percent in the center and windward portions of cities on days with inversions and light winds. In many cities with a naturally humid situation, fogs are much more frequent than in nearby rural areas. In London a change from the inefficient burning of soft coal for domestic heating reduced the frequency of fogs by 50 percent and increased the duration and intensity of the sun's radiation correspondingly.

As the size of metropolitan areas and the total amount of pollution entering the atmosphere from them have increased, the urban influences on the weather have become more extensive. Urbanization has a meteorological significance as well as a geographical and social meaning.

## 15.2   Influences on worldwide weather

In addition to the local influences in places where human activities and the pollution associated with them are highly concentrated, there are global effects due to the combustion of fuels and other industrial processes. The effect that is best established is the increase in the average atmospheric concentration of carbon dioxide during the past few decades. The evidence that shows this most clearly is the record of carbon dioxide concentrations measured at the Mauna Loa Observatory, at an elevation of 11,150 ft on the island of Hawaii. Since this observatory is remote from local sources and sinks of carbon dioxide, the concentrations measured there are representative of the well-mixed atmospheric background.

Figure 15.1 shows the variation in the carbon dioxide concentration at Mauna Loa since 1958. The concentration has been increasing at the average rate of 0.77 parts per million (ppm) per year. This rate of increase is corroborated by measurements at the South Pole and at other places where the observations are not affected by local influences. This accumulation of carbon dioxide in the atmosphere has been compared with the rate of its production by the burning of fossil fuel. This comparison showed that approximately one-half of the carbon dioxide produced from this source

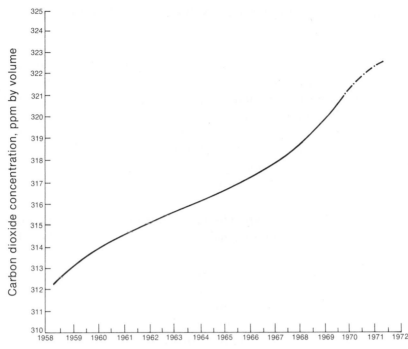

**Figure 15.1** *Measured carbon dioxide concentration at Mauna Loa Observatory. [Adapted from L. Machta, "Mauna Loa and Global Trends in Air Quality," Bull. Am. Meteorol. Soc. 53 (May 1972):402–420.]*

remains in the atmosphere. The remainder is removed either by dissolving in the ocean or by growth in excess of decay of vegetation. Based on projections of the amount of combustion of coal, petroleum, and natural gas that will take place to meet the growing demand for energy, it has been estimated that the carbon dioxide concentration will reach 380 ppm in the year 2000.

Whether human activites have led to a similar increase in the particulate content of the atmosphere has not yet been determined definitely. Direct measurements of particle counts in places away from sources of pollution are not available. As an indirect indication, the transmission of solar radiation through the atmosphere has been used as a measure of the *turbidity*, that is, the amount of particulate matter in the entire path of sunlight through the atmosphere. The measurements are affected by the introduction of large amounts of dust into the stratosphere by large volcanic eruptions, such as the one that occurred at Mt. Agung, on the island of Bali, in 1963. Allowing for the effects of volcanic activity, the records of transmission of solar radiation at Mauna Loa show no tendency for the concentration of particulate matter to increase. In contrast, measurements at Washington, D.C., Davos, Switzerland, and several cities in the Soviet Union suggest that there has been an increase of turbidity in recent years.

Most of these locations are subject to local pollution, but Davos, for instance, is a high-altitude station at some distance from any metropolitan or industrial center.

The fact that the average particulate concentration is not increasing at the Mauna Loa Observatory implies that the particles entering the atmosphere from human activities are removed from the air before it reaches there. The removal processes are of two types: "dry" processes in which clouds and rain are not involved, and "wet" processes in which they are. The dry processes are *fallout,* the settling out of the particles because of gravity, and *impaction,* in which the particles are collected by leaves of trees, wires, buildings, or the membranes of people's eyes and respiratory systems, as they are carried by the wind, or by vehicles or other objects moving through the air. The larger particles are removed rapidly by the dry processes, but the very small ones are carried upward by the turbulent air motions as rapidly as they tend to fall and are swept around objects with the air flow. The removal of the smaller particles is almost entirely due to the wet processes: *rainout,* in which the particle is incorporated into cloud drops, which then grow and fall as precipitation; and *washout,* in which the particles are collected by the raindrops or snowflakes falling from cloud to ground. The length of time that the small particles coming from human activities remain in the air thus depends on the time that elapses before the air into which they are emitted is incorporated into a cloud or a precipitation area. The average period is not known, but it appears from the observations referred to above that the tropospheric air reaching Mauna Loa Observatory has always passed through a cleansing process in or below clouds in crossing the Pacific Ocean, and the air reaching the other stations frequently travels from the pollution sources without doing so. From these considerations one may conclude that in areas remote from sources the particulate concentration will not increase, but at places that can be reached by polluted air before the air is cleansed by rainout or washout any increase in the amount of emissions will be reflected by an increase in the amount of particulate matter in the air.

The influence of the changes in concentration of carbon dioxide and particulate matter on the weather is even more uncertain. Theory predicts that if the carbon dioxide content of the atmosphere increases, there should be an increase in the earth's temperature because of the augmentation of the greenhouse effect (see Section 3.7). When the increase of carbon dioxide content was first recognized, in 1940, the trend of the temperature was examined. As expected, it was found that the average world temperature, as well as could be determined, had increased during the previous half century. Since then, however, the mean temperature of the earth has been decreasing in spite of the continued increase in carbon dioxide concentration. To explain this, the effect of the increase in turbidity has been used. This explanation is plausible, particularly if the particles are present in the stratosphere, into which large volcanic eruptions inject them. The particles reduce the amount of solar radiation reaching the ground, and thereby produce a cooling tendency.

Theoretical computations have been carried out in the attempt to predict the amount of temperature change that would result from a given increase in the carbon dioxide concentration or the amount of particulate matter in the atmosphere. These computations have involved assumptions and simplifications that affect the magnitude of the predicted temperature change. For instance, if it were assumed that an increase in sea surface temperature would result in higher humidities without an increase in total cloudiness, the greenhouse effect of the carbon dioxide increase would be magnified. It would be as though a thicker water-vapor blanket were added to the augmented carbon dioxide blanket in preventing heat from escaping. On the other hand, if the increase in humidity produced more cloudiness, the incoming solar radiation would be reduced and the net temperature increase would be smaller. Depending on the assumptions that are made, estimates of the effect of a doubling of the carbon dioxide content range from 1.3°C to 5°C. The 20 percent increase expected by the year 2000 might be expected to cause a temperature increase between one-fourth and one-half degree. Although this amount of increase would cause glaciers and the polar ice caps to recede, it would not produce a cataclysmic rise in sea level. However, if the carbon dioxide content continued to increase, in another half century the doubled value would be attained and the temperature increase might have disastrous consequences.

Quantitative estimates of the effect of changes in the concentration of particulate matter on the earth's average temperature have not been made. Even the direction of the change has been questioned, for it has been pointed out that although its effect will be to reduce the direct solar radiation reaching the ground if the sky is clear, if there is particulate matter above the clouds over the one-half of the earth that is covered with cloud at any one time, the amount of radiation reflected back to space from the cloud tops will be reduced.

The main point to be made with respect to particulate matter is that volcanic eruptions appear to have much more influence than human activities, both because they put more into the air and because it goes into the stratosphere where it is immune from wet removal processes and can stay for months and years. We must conclude that except for the accidental intervention of occasional large volcanic eruptions the temperature of the earth will continually rise because of the increase in carbon dioxide content of the atmosphere.

## 15.3 Intentional weather modification

The possibility of purposefully influencing the weather has attracted man from earliest times. In ancient times the attempts to influence it consisted of prayers and sacrifices to placate the gods. Only very recently have proposals or attempts having a scientific basis been made.

Modification of weather or climate on all scales, from the general atmospheric circulation to individual cumulus clouds and to the temperature of a vegetable patch, has been proposed. However, actual attempts have been confined almost completely to the smaller-scale phenomena.

Proposals to modify the general circulation have included changing the radiational budget of the earth by introducing clouds of soot or ice crystals into the atmosphere at particular latitudes and heights, or by coating the snow and ice fields at high latitudes with carbon black, or by eliminating the ice from the Arctic Ocean by pumping warmer water from the Atlantic Ocean into it. Because of the uncertainty of the consequences and the tremendous investment of energy, materials, and money that would be required to carry them out, large-scale programs of this type have remained the subject of vague speculations. The largest meteorological feature for which systematic weather modification experiments have been conducted is the hurricane.

At the other end of the scale, in the control of the microclimate of plants, control measures have been part of standard agricultural practice for many years. Crops are protected successfully from wind damage through the use of rows of trees as windbreaks and from frost by means of orchard heaters or wind machines. In addition to these generally used procedures, there have been less conventional attempts — for instance, covering the ground with carbon black to alter the local radiational balance and improve crop growth.

The controls of clouds and precipitation is the aspect of weather modification that has received the most attention in recent years. It received its initial impetus with the discovery by Vincent Schaefer in 1946 that dry ice pellets dropped through a supercooled cloud could initiate the three-phase (Bergeron) process and the finding by Bernard Vonnegut that silver iodide was an effective ice-nucleating agent (see Section 8.5). Since then there have been many tests of cloud seeding to dissipate fog, increase or redistribute precipitation, suppress hail, reduce lightning strokes, which sometimes cause forest fires, or decrease the destructive winds accompanying thunderstorms and hurricanes. In addition to the use of ice-nucleating substances to initiate or modify the three-phase precipitation process in supercooled clouds, there have been experiments attempting to initiate or modify the precipitation process in warm clouds by seeding with salt particles or water spray.

The formation and structure of clouds and the occurrence, form, and amount of precipitation can be influenced in two ways: (1) by altering the dynamic processes, that is, the air currents producing the clouds (see Chapter 9); and (2) by changing the microphysical processes of formation and growth of cloud and precipitation particles (see Chapter 8). To alter the air flow directly would require very large amounts of energy. However modifying the microphysical processes in some circumstances produces large changes in the dynamic processes. In turn, changes in the dynamic processes will alter the course of the microphysical processes. The total effect on the presence of cloud and the amount of precipitation depends on the interaction of the two types of processes. Thus, in most weather modification efforts materials (seeding agents) are introduced into clouds to change the growth processes of the cloud particles. This may be done by flying an airplane above the cloud and dropping dry-ice pellets or pyrotechnic devices that release large quantities of silver iodide as they fall, by releasing silver iodide smoke or water spray as a plane passes through the cloud, or by releasing silver iodide smoke or salt particles from ground-based generators and depending on the updrafts that are producing the cloud to carry the seeding agent into the cloud.

The most frequent purpose of cloud seeding is to increase the amount of precipitation in order to increase the crop yield in arid or semiarid regions or regions experiencing a drought, to augment the water supply for domestic or commercial purposes, or to increase the flow of water for hydroelectric power generation. There have been many "rainmaking" projects in various parts of the world. Initially, they were undertaken with a great deal of optimism. Large precipitation increases were attributed to the cloud seeding, particularly by commercial firms, which claimed to have produced increases of several hundred percent. With the passage of time these claims have been modified, and the advocates of cloud seeding usually estimate the average amount of increase to be between 10 and 20 percent.

The reason for uncertainty about the effect of cloud seeding on the amount of precipitation is that if one seeds a cloud one cannot know how much rain would have fallen from it if it had not been seeded. To evaluate the success of cloud-seeding projects comparisons have been made with control areas—nearby areas that have not been seeded but, according to previous records, have rain amounts correlated with the natural precipitation amounts in the target area. This method of evaluation has been severely criticized by statisticians. To overcome these criticisms experiments have been designed employing randomization. In a randomized experiment the situations that are considered suitable for seeding are divided into two groups on the basis of pure chance, equivalent to tossing a coin. One of the groups is seeded, and the other is not. The difference in precipitation between the seeded group and the unseeded group is attributed to seeding, the assumption being that the effects of all other factors cancel one another because of randomization.

A number of randomized experiments have been carried out, with the surprising result that although in some of them the seeded group had more precipitation than the unseeded, in many of them there was no significant

A

B

**Figure 15.2**  *Results of seeding stratocumulus clouds. Dry ice was dropped into the clouds at the rate of 12 lb/mi along three 9-mile-long lines three miles apart. In photograph A, taken 15 minutes after seeding, the change in the character of the cloud top is just beginning to be evident. Photograph B, taken 35 minutes after seeding, shows a marked change in the character of the cloud. The ground could be seen from above*

*40 minutes after seeding, and 75 minutes after seeding, photograph C, small cumulus clouds had formed in the cleared area. Photograph D, taken 80 minutes after seeding when the plane had descended through the cleared area, shows snow showers falling from the clouds in the distance. [Courtesy of U.S. Army.]*

difference and in some there was significantly less precipitation for the seeded group than for the unseeded group. For example, a randomized experiment, designated Project Whitetop, was conducted in Missouri during the summers of 1960–1964 to test the precipitation effects of seeding summer cumulus with silver iodide by airplane. The days on which to seed were selected by using objective criteria that were designed to identify days on which seeding would be expected to augment the amount of rain. Analysis of the results showed definitely smaller amounts of precipitation on seeded days than on unseeded days. On the other hand, an experiment conducted in Israel from 1961 to 1966, in which winter storm situations were selected for seeding, indicated an increase in precipitation of 18 percent due to seeding.

Actually, this result, that some of the experiments showed smaller amounts of precipitation with seeding, should not surprise us, for it is easy to understand why seeding should have a precipitation-reducing effect under some circumstances. For example, if the number of natural ice nuclei happened to be optimal for the production of precipitation, the addition of artificial nuclei would lead to larger numbers of smaller ice crystals. Overseeding in this fashion has in fact been suggested as a method of combating flood-producing rains. As another example, consider the effect of seeding cumulus clouds. In convective situations rain occurs naturally after the cloud top has risen far enough to reach temperatures at which the nuclei present are effective. If the cloud is seeded before it reaches this level, the precipitation will start when the cloud is not as deep, and further growth of the cloud will be inhibited. The precipitation from the shallower cloud may be considerably less than that which would have occurred had it grown to its full depth before precipitation began. These examples, overseeding and premature development of precipitation, are just two of several ways in which seeding might lead to a reduction in the amount of precipitation.

On the other hand, in some situations the convection is limited to the layer below a slight inversion and the release of the latent heat of freezing when the cloud is seeded gives just enough added buoyancy to the cloud top to enable the cloud to penetrate the inversion. If this occurs, the cloud may grow explosively and a large amount of rain may fall from a cloud that would not have precipitated at all in the absence of seeding.

We conclude that there are some circumstances in which cloud seeding can increase precipitation and some circumstances in which it will decrease precipitation. At present we have only a partial understanding of what these circumstances are. The objective of much of the cloud-physics research currently underway is to find out how to tell in which situations cloud seeding will increase the amount to precipitation.

One of the approaches to determining whether or not a situation would yield more precipitation with seeding than without is numerical modelling.

Ideally, given the observations of the general meteorological situation and the measurements of condensation and ice nuclei, it should be possible to compute the rate of development of cloud and precipitation. Then the computation could be repeated with artificial nuclei added, and the result would indicate whether seeding would lead to more precipitation. Considerable progress has been made in developing numerical models that deal with one or another aspect of the overall problem, but so far the models are highly simplified. In spite of the simplicity of the models, they have been applied with some success to the selection of situations suitable for seeding.

Most of the attempts to augment precipitation have been based on the three-phase process and the use of dry ice or silver iodide as seeding agents. There have also been a few experiments to test the possibility of stimulating the warm-rain (collision-coalescence) process. Experiments in Australia, over the Caribbean, and in the central United States have shown a tendency for cumulus clouds to develop precipitation when water was sprayed into their bases. However, the cost of transporting the large quantities of water needed by airplane and the relatively small increase in precipitation make the procedure questionable from the economic standpoint. Seeding by using giant hygroscopic nuclei appears to have more potential for warm clouds. The seeding can be carried out from the ground, greatly reducing the cost. Experiments conducted in India and Pakistan have reported success in producing modest increases in precipitation with this method.

The dissipation of supercooled fog is perhaps the most successful result of cloud seeding. It has become an operational practice in airports in various parts of the world. The procedure used at some cities in the United States that are subject to frequent fogs in winter is to alert the pilots of small seeding planes when fog is expected, so that they can take off before the fog closes the airport. Dry-ice pellets are released from the seeding plane into the fog at a suitable distance upwind of the runway; then the pilot waits for the development of a hole in the fog as the ice crystals grow and fall out. Unless the wind carries new fog onto the runway, making additional seeding necessary, the plane lands on the cleared runway, which is then available for landing and takeoff by other planes. At the Orly airport near Paris, liquid propane released through expansion orifices is used to seed the clouds. The cooling by expansion as the propane evaporates produces the ice crystals. This procedure has the advantage that it is carried on with equipment at the ground.

From the standpoint of airport operations, the dissipation of warm fog is more important than that of supercooled fog, because a larger portion of the hours of low visibility occurs at temperatures above 0°C at most of the busy airports of the world. Although seeding with giant hygroscopic nuclei or polyelectrolytes and flying helicopters across the top of the cloud to mix warm dry air into the foggy air in the downwash of the rotors have had some limited success, none of these procedures has yet shown enough reliability to be put into routine use.

Hail suppression for the protection of agricultural crops is another objective of weather modification that has received much attention. Attempts to prevent hail were made long before the introduction of seeding with dry ice or silver iodide. For many years, gunfire and rockets were used for this purpose in France, Italy, and Switzerland. Success has been claimed for some of these trials, although no sound explanations why they should work have been put forward. On the other hand, for seeding with silver iodide a reasonable basis has been proposed, namely, that the conversion of the cloud to ice early in its development will prevent the growth of hailstones. There have been tests of seeding to reduce hail in several countries with variable results. The ones reported to have had the greatest success have been a program in France, in which seeding from 1959 to 1966 produced a decrease of 22.6 percent in annual loss due to hail damage, compared with the previous 15 years, and large-scale hail prevention programs in the Soviet Union, in which reductions of damage of 50 to 90 percent in protected areas, compared with areas that were not protected, were claimed. In the Russian procedure, the seeding is accomplished by rockets or artillery shells that are directed by radar to the precise position in the storm cloud at which the hail production is taking place. Refrangible shells and rockets were developed, which break up into very small fragments when they explode so that the debris falling from the spent projectiles can do no harm. The claims of success of the Russian method have stimulated a large-scale research program in the United States to test the theories on which their method is based. The study of the structure of hailstorms and of the effects of seeding them is expected to lead to improved methods of hail control in this country.

As lightning is the principal cause of forest fires in the extensive forests of the United States, the possibility that its occurrence could be reduced by seeding has led to extensive investigations conducted by the U.S. Forest Service and by some state divisions of forestry. There has been some indication that seeding thunderstorms with silver iodide does reduce the number of lightning strokes, but the results are not yet conclusive. Another method that has been tried utilizes chaff — metal-covered nylon needles — which tends to discharge the cloud before the electric potential required for lightning can develop. Tests suggest that the method would be successful if a practical means of dispersing the chaff throughout the cloud could be developed.

In hurricane modification experiments the purpose is to reduce the strength of the destructive winds near the center. The horizontal convergence in hurricanes is concentrated at the eye wall and along the spiral bands radiating out from it. The highest clouds and heaviest precipitation occur there. Associated with the extremely rapid release of latent heat, the strongest winds occur in a ring around the eye wall. Attempts to reduce their severity are based on the idea that broadening the area over which the latent heat is released will increase the distance over which the pressure decreases, reducing the pressure gradient and thus weakening the winds. A few experiments have been conducted in which massive amounts of silver iodide were released into the eye wall of hurricanes by dropping pyrotechnic devices from planes. The winds were lower for a while after the seeding in these experiments, but by amounts that are within the normal variability of winds in hurricanes. After a time they increased again. Since randomized experiments with hurricanes cannot be conducted because there are so few of them, the possibility that the decreases in wind were not due to the seeding cannot be rejected. However, if the same result is obtained in future tests, the probability will be reinforced that seeding is producing the reduction in wind speed.

In reviewing the various weather modification attempts discussed herein, we see that the potential for weather modification exists but the observational data and the theoretical understanding of cloud and precipitation physics are not adequate to enable us to predict with certainty the effect of cloud seeding. Much research remains to be done before we can truly control the weather.

We can visualize the ultimate state of the weather-control process as part of an "ideal" weather service. The data from satellites, balloons, and surface weather stations would be fed automatically into computers, which would process them to forecast the conditions that would develop if no modification treatment is applied. On the basis of criteria for social, economic, and aesthetic benefits, the treatment that would be required to produce optimal weather conditions for the largest number of people would be arrived at by the computer, which would then automatically turn on the nucleus generators and other devices to administer the treatment needed.

We are far from such a state, and may never achieve it. But just like the outcome of improvements in forecasting, the social good and economic benefits that would result from success in weather control justify increased efforts to develop the scientific basis for them, namely, a fuller understanding of our atmospheric environment.

## Questions, Problems, and Projects for Chapter 15

1 What do you think were the changes in the climate of the United States caused by the settlement by the Europeans, with accompanying spread of agriculture and development of large cities?

2 What would happen to the concentration of carbon dioxide in the atmosphere

if nuclear reactors completely replaced combustion of fossil fuels as a source of energy? What would be the effect on the temperature of the earth?

3  Compare the processes that lead to the removal of particulates, as listed in Section 15.2, with those that remove carbon dioxide from the atmosphere. Why is the removal of particulates virtually complete, while about one-half of the carbon dioxide produced by combustion remains in the air? Discuss the processes you consider effective in removing sulfur dioxide, oxides of nitrogen, and other gaseous contaminants.

4  Explain why cloud seeding with dry ice or silver iodide may be expected to increase precipitation. Why does it sometimes decrease precipitation?

5  Why does the introduction of water spray or giant hygroscopic nuclei into warm clouds tend to initiate or increase precipitation? Are there circumstances in which it might decrease precipitation?

6  Discuss the problem of testing the success of a "rainmaking" experiment.

# Appendix

## A  Units of measurement and conversion factors

The *metric system* is used generally in science, and in commerce and everyday life everywhere except in some English-speaking countries. It is based on the *meter* for length and the second for time; the unit of mass, the kilogram, and of volume, the liter, are defined in terms of the meter, and units of force (dyne, newton), pressure (bar), energy (erg, joule), power (watt), and so forth, are defined in terms of the meter, kilogram, and second.

Multiples or fractions of units, in powers of ten, are designated by prefixes as follows:

| | | | | |
|---|---|---|---|---|
| p | pico- | one-trillionth (U.S.) | $10^{-12}$ | 0.000 000 000 001 |
| n | nano- | one-billionth (U.S.) | $10^{-9}$ | 0.000 000 001 |
| $\mu$ | micro- | one-millionth | $10^{-6}$ | 0.000 001 |
| m | milli- | one-thousandth | $10^{-3}$ | 0.001 |
| c | centi- | one-hundredth | $10^{-2}$ | 0.01 |
| d | deci- | one-tenth | $10^{-1}$ | 0.1 |
| da | deka- | ten | $10$ | 10. |
| h | hecto- | one hundred | $10^{2}$ | 100. |
| k | kilo- | one thousand | $10^{3}$ | 1 000. |

M  mega-  one million        $10^6$       1 000 000.
G  giga-  one billion (U.S.)  $10^9$       1 000 000 000.
T  tera-  one trillion (U.S.)  $10^{12}$  1 000 000 000 000.

## Table of Units and Equivalents

| Metric units | English units | Conversions |
|---|---|---|
| **LENGTH** | | |
| centimeter (cm) | inch (in) | 1 cm = 0.3937 in; 1 in = 2.54 cm |
| meter (m) | foot (ft) | 1 m = 3.28 ft; 1 ft = 0.305 m |
| kilometer (km) | mile (mi) | 1 km = 0.621 mi; 1 mi = 1.61 km |
| micron ($\mu$m) | | 1 $\mu$m = $10^{-6}$ m = $10^{-4}$ cm = 3.94 · $10^{-5}$ in |
| Ångstrom unit (Å) | | 1 Å = $10^{-8}$ cm = $10^{-4}$ $\mu$m |

One degree of latitude = 111.1 km = 69.1 mi = 60 nautical miles
One nautical mile = 1.15 statute miles = 1.85 km

| | | |
|---|---|---|
| **MASS** | | |
| gram (g) | ounce (oz av) | 1 g = 0.0353 oz; 1 oz = 28.35 g |
| kilogram (kg) | pound (lb) | 1 kg = 2.20 lbs; 1 lb = 0.454 kg |
| metric ton (m ton = $10^3$ kg) | short ton (2000 lbs) | 1 m ton = 1.10 short tons <br> 1 short ton = 0.907 m tons |

| | | |
|---|---|---|
| **TEMPERATURE** | | |
| Celsius degree (C°) | Fahrenheit degree (F°) | 1 C° = 1.8 F°; 1 F° = 5/9 C° |

| | | |
|---|---|---|
| **CONVERSIONS OF TEMPERATURE** | | |
| Celsius or centigrade C ($T_C$) | | $T_C = 5/9 \, (T_F - 32)$ °C |
| Kelvin or Absolute K ($T$) | Fahrenheit F ($T_F$) | $T_F = [1.8 \, T_C + 32]$ °F <br> $T = [T_C + 273.15]$ °K |

| | | |
|---|---|---|
| **AREA** | | |
| square centimeter (cm²) | square inch (sq in) | 1 cm² = 0.155 sq in <br> 1 sq in = 6.45 cm² |
| square meter (m²) | square foot (sq ft) | 1 m² = 10.76 sq ft <br> 1 sq ft = 0.093 m² |
| square kilometer (km²) | square mile (sq mi) | 1 km² = 0.386 sq mi <br> 1 sq mi = 2.59 km² |
| hectare | acre | 1 hectare = 10,000 m² = 2.47 acres <br> 1 acre = 43,560 sq ft = 0.4047 hectares |

| | | |
|---|---|---|
| **VOLUME** | | |
| cubic centimeter (cm³) | cubic inch (cu in) | 1 cm³ = 0.061 cu in <br> 1 cu in = 16.39 cm³ |
| cubic meter (m³) | cubic foot (cu ft) | 1 m³ = 35.3 cu ft <br> 1 cu ft = 0.028 m³ |

| liter | gallon | 1 liter = 1,000 cm³ = 0.264 gallons<br>1 gallon = 231 cu in = 3.7853 liters |
|---|---|---|

SPEED

| centimeter per second (cm s⁻¹) | foot per second (ft/sec) | 1 cm s⁻¹ = 0.0328 ft/sec = 1.97 ft/min<br>1 ft/sec = 30.48 cm s⁻¹ = 0.592 kt |
|---|---|---|
| meter per second (m s⁻¹) | mile per hour (mi/hr) | 1 m s⁻¹ = 2.24 mi/hr = 1.94 kt<br>1 mi/hr = 0.447 m s⁻¹ = 0.868 kt |

knot (kt)   1 kt = 1 nautical mile per hour = 0.515 m s⁻¹ = 1.15 mi/hr

WORK AND ENERGY

| erg | | 1 joule = $10^7$ ergs<br>1 cal = 4.186 · $10^7$ ergs = 4.186 joules |
|---|---|---|
| joule (J) | foot pound (ft lb) | 1 ft lb = 1.36 joules<br>1 joule = 0.738 ft lb |
| calorie (cal) | British Thermal Unit (BTU) | 1 BTU = 252 cal = 1055 joules<br>1 cal = 3.97 · $10^{-3}$ BTU |
| langley (ly) | | 1 ly = 1 cal/cm² = 4.186 · $10^4$ J/m² |
| watt (W) | horsepower (hp) | 1 W = 1 J/sec = 1.34 · $10^{-3}$ hp<br>1 hp = 33,000 ft lb/min = 746 W<br>1 ly/min = 6.98 · $10^2$ W/m² |

PRESSURE

| dyne/cm² | lb/sq in | 1 dyne/cm² = 1.45 · $10^{-5}$ lb/sq in<br>1 lb/sq in = 6.89 · $10^4$ dyne/cm² |
|---|---|---|
| millibar (mb) | | 1 mb = $10^3$ dyne/cm² = 0.750 mm Hg = 0.0295 in Hg |
| millimeters of mercury (mm Hg) | inches of mercury (in Hg) | 1 mm Hg = 1.333 mb = 0.03937 in Hg = 0.0193 lb/sq in<br>1 in Hg = 33.86 mb = 25.4 mm Hg = 0.491 lb/sq in |

One atmosphere = 1013.25 mb = 760 mm Hg = 29.92 in Hg = 14.7 lb/sq in

---

## B   Numerical constants

| | |
|---|---|
| Mean solar distance | 1.495 · $10^8$ km |
| Equatorial radius of earth | 6378.4 km |
| Polar radius of earth | 6356.9 km |
| Angular velocity of earth's rotation | $\Omega = 7.29 \cdot 10^{-5}$ s⁻¹ |
| Acceleration of gravity (at sea level and 45° lat) | 9.806 m s⁻² |
| Solar constant | 1.94 ly/min = 1.35 kW/m² |

| | |
|---|---|
| Gas constant for dry air | 287 J/(kg °K) |
| Mean molecular weight of dry air | 29.0 g/mole |
| Specific heat of dry air: | |
| at constant pressure $c_p$ | 1003 J/(kg °K) |
| at constant volume $c_v$ | 717 J/(kg °K) |

## C  Greek alphabet

In science the number of quantities that must be represented by symbols in equations is so large that in addition to the Latin alphabet used in English, letters from other alphabets are used — particularly, the Greek. The following list gives the Greek letters, many of which are commonly used in meteorology. The capital letters that are identical with Latin letters are shown in parenthesis.

| Small Letter | Capital Letter | Name | Small Letter | Capital Letter | Name |
|---|---|---|---|---|---|
| $\alpha$ | (A) | alpha | $\nu$ | (N) | nu |
| $\beta$ | (B) | beta | $\xi$ | $\Xi$ | xi |
| $\gamma$ | $\Gamma$ | gamma | $o$ | (O) | omicron |
| $\delta$ | $\Delta$ | delta | $\pi$ | $\Pi$ | pi |
| $\epsilon$ | (E) | epsilon | $\rho$ | (P) | rho |
| $\zeta$ | (Z) | zeta | $\sigma$ | $\Sigma$ | sigma |
| $\eta$ | (H) | eta | $\tau$ | (T) | tau |
| $\theta$ | $\Theta$ | theta | $\upsilon$ | $Y$ | upsilon |
| $\iota$ | (I) | iota | $\phi$ | $\Phi$ | phi |
| $\kappa$ | (K) | kappa | $\chi$ | (X) | chi |
| $\lambda$ | $\Lambda$ | lambda | $\psi$ | $\Psi$ | psi |
| $\mu$ | (M) | mu | $\omega$ | $\Omega$ | omega |

## D  Weather-map symbols

The symbols that are used on surface weather maps to summarize the conditions observed at individual stations are given on the following pages.[1] The data are transmitted by the observation station to the collection center by teletype or radio in the form of a coded message. They are then plotted around a small circle in the position on the map where the station is located, using the arrangement shown in the station model and the symbols shown in the various tables.

---

[1]The symbols and map entries were prescribed by the World Meteorological Organization for use by the meteorological services of all countries. The presentation in this appendix is taken from the U.S. National Weather Service publication "Explanation of the Weather Map."

**1**

## SYMBOLIC STATION MODEL

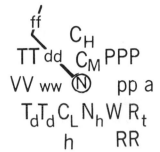

## SAMPLE PLOTTED REPORT

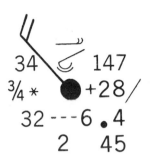

**2**

## EXPLANATION OF SYMBOLS AND MAP ENTRIES

| Symbols in order as they appear in the message | Explanation of symbols and decode of sample message in block | Remarks on coding and plotting |
|---|---|---|
| Iliii | Station number $72405 =$ Washington | Usually printed on manuscript maps below station circle. Omitted on Daily Weather Map in favor of printed station names. |
| N | Total amount of cloud $8 =$ completely covered | Observed in tenths of cloud cover and coded according to code table in block **5** Plotted in symbols shown in same table. |
| dd | True direction from which wind is blowing $32 = 320° =$ NW | Coded in tens of degrees and plotted as the shaft of an arrow extending from the station circle toward the direction from which the wind is blowing. |
| ff | Wind speed in knots $20 = 20$ knots | Coded in knots (nautical miles per hour) and plotted as feathers and half-feathers representing 10 and 5 knots, respectively, on the shaft of the wind direction arrow. See block **9**. |
| VV | Visibility in miles and fractions $12 = 12/16$ or $3/4$ miles | Decoded and plotted in miles and fractions up to $3\frac{1}{8}$ miles. Visibilities above $3\frac{1}{8}$ miles but less than 10 miles are plotted to the nearest whole mile. Values higher than 10 miles are omitted from the map. |
| ww | Present weather $70 =$ slight intermittent snow | Coded in figures taken from the "ww" table (block **8** ) and plotted in the corresponding symbols same block. Entries for code figures 00, 01, 02, and 03 are omitted from this map. |
| W | Past weather $6 =$ rain | Coded in figures taken from the "W" table (block **11** ) and plotted in the corresponding symbols same block. No entry made for code figures 0, 1, or 2 |

**2** (continued)

| | | |
|---|---|---|
| PPP | Barometric Pressure (in millibars) reduced to sea-level 147 = 1014.7 mb. | Code and plotted in tens, units, and tenths of millibars. The initial 9 or 10 and the decimal point are omitted. |
| TT | Current air temperature in °C. 01 = 34°F. | Coded in whole degrees Celsius and plotted in nearest equivalent whole degrees Fahrenheit. |
| $N_h$ | Fraction of sky covered by low or middle cloud 6 = 7 or 8 tenths | Observed in tenths of cloud cover and coded according to code table in block **6**. Plotted on map as code figure in message. |
| $C_L$ | Cloud type 7 = Fractostratus and/or Fractocumulus of bad weather (scud) | Predominating clouds of types in $C_L$ table (block **3**) are coded from that table and plotted in corresponding symbols. "0" is coded if no $C_L$ clouds are observed. |
| h | Height above ground, of base of lowest cloud. 2 = 300 to 599 feet | Observed in feet and coded and plotted as code figures according to code table in block **7**. |
| $C_M$ | Cloud type 9 = Altocumulus of chaotic sky | See $C_M$ table block **3**. |
| $C_H$ | Cloud type 2 = Dense cirrus in patches | See $C_H$ table in block **3**. |
| $T_d T_d$ | Temperature of dewpoint in °C. 00 = 32°F. | Same as TT above. |
| a | Characteristic of barograph trace 2 = rising steadily or unsteadily | Coded according to table in block **10** and plotted in corresponding symbols. |
| pp | Pressure change in 3 hours preceding observation 28 = 2.8 millibars | Coded and plotted in units and tenths of millibars. |
| 7 | Indicator figure | Not plotted. |
| RR | Amount of precipitation 45 = 0.45 inches | Coded and plotted in inches to the nearest hundredth of an inch. If precipitation less than .005 inch is reported, RR is coded 00 and plotted as T |
| $R_t$ | Time precipitation began or ended 4 = 3 to 4 hours ago | Coded and plotted in figures from table in block **4**. |
| s | Depth of snow on ground in inches 1=1 inch | Not plotted. (used in Snowdepth analysis) |

**3**

| Code No. $C_L$ | DESCRIPTION (Abridged From International Code) |
|---|---|
| 1 | Cu of fair weather, little vertical development and seemingly flattened |
| 2 | Cu of considerable development, generally towering, with or without other Cu or Sc bases all at same level |
| 3 | Cb with tops lacking clear-cut outlines, but distinctly not cirriform or anvil-shaped; with or without Cu, Sc, or St |
| 4 | Sc formed by spreading out of Cu; Cu often present also |
| 5 | Sc not formed by spreading out of Cu |
| 6 | St or Fs or both, but no Fs of bad weather |
| 7 | Fs and/or Fc of bad weather (scud) |
| 8 | Cu and Sc (not formed by spreading out of Cu) with bases at different levels |
| 9 | Cb having a clearly fibrous (cirriform) top, often anvil-shaped, with or without Cu, Sc, St, or scud |

| Code No. $C_H$ | DESCRIPTION (Abridged From International Code) |
|---|---|
| 1 | Filaments of Ci, or "mares tails," scattered and not increasing |
| 2 | Dense Ci in patches or twisted sheaves, usually not increasing, sometimes like remains of Cb; or towers or tufts |
| 3 | Dense Ci, often anvil-shaped, derived from or associated with Cb |
| 4 | Ci, often hook-shaped, gradually spreading over the sky and usually thickening as a whole |
| 5 | Ci and Cs, often in converging bands, or Cs alone; generally overspreading and growing denser; the continuous layer not reaching 45° altitude |
| 6 | Ci and Cs, often in converging bands, or Cs alone; generally overspreading and growing denser; the continuous layer exceeding 45° altitude |
| 7 | Veil of Cs covering the entire sky |
| 8 | Cs not increasing and not covering entire sky |
| 9 | Cc alone or Cc with some Ci or Cs, but the Cc being the main cirriform cloud |

| Code No. $C_M$ | DESCRIPTION (Abridged From International Code) |
|---|---|
| 1 | Thin As (most of cloud layer semi-transparent) |
| 2 | Thick As, greater part sufficiently dense to hide sun (or moon), or Ns |
| 3 | Thin Ac, mostly semi-transparent; cloud elements not changing much and at a single level |
| 4 | Thin Ac in patches; cloud elements continually changing and/or occurring at more than one level |
| 5 | Thin Ac in bands or in a layer gradually spreading over sky and usually thickening as a whole |
| 6 | Ac formed by the spreading out of Cu or Cb |
| 7 | Double-layered Ac, or a thick layer of Ac, not increasing; or Ac with As and/or Ns |
| 8 | Ac in the form of Cu-shaped tufts or Ac with turrets |
| 9 | Ac of a chaotic sky, usually at different levels; patches of dense Ci are usually present also |

**4**

R$_t$ **TIME PRECIPITATION BEGAN OR ENDED**

0    No Precipitation

1    Less than 1 hour ago

2    1 to 2 hours ago

3    2 to 3 hours ago

4    3 to 4 hours ago

5    4 to 5 hours ago

6    5 to 6 hours ago

7    6 to 12 hours ago

8    More than 12 hours ago

9    Unknown

**5**

**Code No.** N **SKY COVERAGE** (Total Amount)

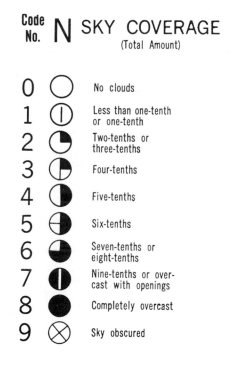

0    No clouds

1    Less than one-tenth or one-tenth

2    Two-tenths or three-tenths

3    Four-tenths

4    Five-tenths

5    Six-tenths

6    Seven-tenths or eight-tenths

7    Nine-tenths or overcast with openings

8    Completely overcast

9    Sky obscured

## 6

# N$h$   SKY COVERAGE
(Low And/Or Middle Clouds)

0   No clouds

1   Less than one-tenth
or one-tenth

2   Two-tenths or
three-tenths

3   Four-tenths

4   Five-tenths

5   Six tenths

6   Seven-tenths or
eight-tenths

7   Nine-tenths or over-
cast with openings

8   Completely overcast

9   Sky obscured

## 7

# h   HEIGHT
IN FEET
(Approximate)

0   0–149

1   150–299

2   300–599

3   600 999

4   1,000–1,999

5   2,000–3,499

6   3,500–4,999

7   5,000–6,499

8   6,500–7,999

9   At or above
8,000, or no
clouds

**8**

# WW PRESENT WEATHER

## (Descriptions Abridged from International Code)

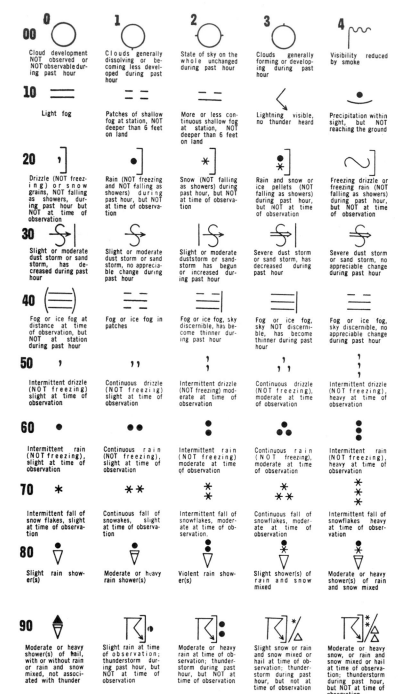

| | **0** | **1** | **2** | **3** | **4** |
|---|---|---|---|---|---|
| **00** | Cloud development NOT observed or NOT observable during past hour | Clouds generally dissolving or becoming less developed during past hour | State of sky on the whole unchanged during past hour | Clouds generally forming or developing during past hour | Visibility reduced by smoke |
| **10** | Light fog | Patches of shallow fog at station, NOT deeper than 6 feet on land | More or less continuous shallow fog at station, NOT deeper than 6 feet on land | Lightning visible, no thunder heard | Precipitation within sight, but NOT reaching the ground |
| **20** | Drizzle (NOT freezing) or snow grains, NOT falling as showers, during past hour but NOT at time of observation | Rain (NOT freezing and NOT falling as showers) during past hour, but NOT at time of observation | Snow (NOT falling as showers) during past hour, but NOT at time of observation | Rain and snow or ice pellets (NOT falling as showers) during past hour, but NOT at time of observation | Freezing drizzle or freezing rain (NOT falling as showers) during past hour, but NOT at time of observation |
| **30** | Slight or moderate dust storm or sand storm, has decreased during past hour | Slight or moderate dust storm or sand storm, no appreciable change during past hour | Slight or moderate duststorm or sandstorm has begun or increased during past hour | Severe dust storm or sand storm, has decreased during past hour | Severe dust storm or sand storm, no appreciable change during past hour |
| **40** | Fog or ice fog at distance at time of observation, but NOT at station during past hour | Fog or ice fog in patches | Fog or ice fog, sky discernible, has become thinner during past hour | Fog or ice fog, sky NOT discernible, has become thinner during past hour | Fog or ice fog, sky discernible, no appreciable change during past hour |
| **50** | Intermittent drizzle (NOT freezing) slight at time of observation | Continuous drizzle (NOT freezing) slight at time of observation | Intermittent drizzle (NOT freezing) moderate at time of observation | Continuous drizzle (NOT freezing), moderate at time of observation | Intermittent drizzle (NOT freezing), heavy at time of observation |
| **60** | Intermittent rain (NOT freezing), slight at time of observation | Continuous rain (NOT freezing), slight at time of observation | Intermittent rain (NOT freezing) moderate at time of observation | Continuous rain (NOT freezing), moderate at time of observation | Intermittent rain (NOT freezing), heavy at time of observation |
| **70** | Intermittent fall of snow flakes, slight at time of observation | Continuous fall of snowakes, slight at time of observation | Intermittent fall of snowflakes, moderate at time of observation | Continuous fall of snowflakes, moderate at time of observation | Intermittent fall of snowflakes heavy at time of observation |
| **80** | Slight rain shower(s) | Moderate or heavy rain shower(s) | Violent rain shower(s) | Slight shower(s) of rain and snow mixed | Moderate or heavy shower(s) of rain and snow mixed |
| **90** | Moderate or heavy shower(s) of hail, with or without rain or rain and snow mixed, not associated with thunder | Slight rain at time of observation; thunderstorm during past hour, but NOT at time of observation | Moderate or heavy rain at time of observation; thunderstorm during past hour, but NOT at time of observation | Slight snow or rain and snow mixed or hail at time of observation; thunderstorm during past hour, but not at time of observation | Moderate or heavy snow, or rain and snow mixed or hail at time of observation; thunderstorm during past hour, but NOT at time of observation. |

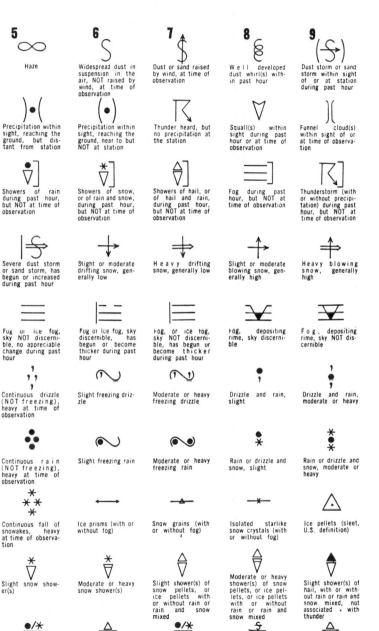

**5** ∞ — Haze

**6** — Widespread dust in suspension in the air, NOT raised by wind, at time of observation

**7** — Dust or sand raised by wind, at time of observation

**8** — Well developed dust whirl(s) within past hour

**9** — Dust storm or sand storm within sight of or during past hour

Precipitation within sight, reaching the ground, but distant from station

Precipitation within sight, reaching the ground, near to but NOT at station

Thunder heard, but no precipitation at the station

Squall(s) within sight during past hour or at time of observation

Funnel cloud(s) within sight of or at time of observation

Showers of rain during past hour, but NOT at time of observation

Showers of snow, or of rain and snow, during past hour, but NOT at time of observation

Showers of hail, or of hail and rain, during past hour, but NOT at time of observation

Fog during past hour, but NOT at time of observation

Thunderstorm (with or without precipitation) during past hour, but NOT at time of observation

Severe dust storm or sand storm, has begun or increased during past hour

Slight or moderate drifting snow, generally low

Heavy drifting snow, generally low

Slight or moderate blowing snow, generally high

Heavy blowing snow, generally high

Fog or ice fog, sky NOT discernible, no appreciable change during past hour

Fog or ice fog, sky discernible, has begun or become thicker during past hour

Fog, or ice fog, sky NOT discernible, has begun or become thicker during past hour

Fog, depositing rime, sky discernible

Fog, depositing rime, sky NOT discernible

Continuous drizzle (NOT freezing), heavy at time of observation

Slight freezing drizzle

Moderate or heavy freezing drizzle

Drizzle and rain, slight

Drizzle and rain, moderate or heavy

Continuous rain (NOT freezing), heavy at time of observation

Slight freezing rain

Moderate or heavy freezing rain

Rain or drizzle and snow, slight

Rain or drizzle and snow, moderate or heavy

Continuous fall of snowakes, heavy at time of observation

Ice prisms (with or without fog)

Snow grains (with or without fog)

Isolated starlike snow crystals (with or without fog)

Ice pellets (sleet, U.S. definition)

Slight snow shower(s)

Moderate or heavy snow shower(s)

Slight shower(s) of snow pellets, or ice pellets with or without rain or rain and snow mixed

Moderate or heavy shower(s) of snow pellets, or ice pellets with or without rain or rain and snow mixed

Slight shower(s) of hail, with or without rain or rain and snow mixed, not associated with thunder

Slight or moderate thunderstorm without hail, but with rain and/or snow at time of observation

Slight or moderate thunderstorm, with hail at time of observation

Heavy thunderstorm, without hail, but with rain and/or snow at time of observation

Thunderstorm combined with dust storm or sand storm at time of observation

Heavy thunderstorm with hail at time of observation

## 9

| ff | KNOTS | (MILES) (Statute) Per Hour |
|---|---|---|
| ◎ | Calm | Calm |
| — | 1–2 | 1–2 |
| | 3–7 | 3–8 |
| | 8–12 | 9–14 |
| | 13–17 | 15–20 |
| | 18–22 | 21–25 |
| | 23–27 | 26–31 |
| | 28–32 | 32–37 |
| | 33–37 | 38–43 |
| | 38–42 | 44–49 |
| | 43–47 | 50–54 |
| | 48–52 | 55–60 |
| | 53–57 | 61–66 |
| | 58–62 | 67–71 |
| | 63–67 | 72–77 |
| | 68–72 | 78–83 |
| | 73–77 | 84–89 |
| | 103–107 | 119–123 |

## 10

### PRESSURE TENDENCY

| Code No. | a | |
|---|---|---|
| 0 | ∧ | Rising, then falling same or higher than 3 hours ago |
| 1 | / | Rising, then steady; or rising, then rising more slowly |
| 2 | / | Rising steadily, or unsteadily |
| 3 | ✓ | Falling or steady, then rising; or rising, then rising more rapidly |
| 4 | — | Steady, same as 3 hours ago |
| 5 | ∨ | Falling, then rising, same or lower than 3 hours ago |
| 6 | \ | Falling, then steady; or falling, then falling more slowly |
| 7 | \ | Falling steadily, or unsteadily |
| 8 | ∧ | Steady or rising, then falling; or falling, then falling more quickly |

Codes 1, 2, 3: Barometric pressure now higher than 3 hours ago

Codes 6, 7, 8: Barometric pressure now lower than 3 hours ago

## 11

### PAST WEATHER

| Code No. | W | |
|---|---|---|
| 0 | | Clear or few clouds |
| 1 | | Partly cloudy (scattered) or variable sky |
| 2 | | Cloudy (broken) or overcast |
| 3 | ⚡/† | Sandstorm or dust-storm, or drifting or blowing snow |
| 4 | ≡ | Fog, ice fog, thick haze or thick smoke |
| 5 | , | Drizzle |
| 6 | • | Rain |
| 7 | ✶ | Snow, or rain and snow mixed, or ice pellets (sleet) |
| 8 | ▽ | Shower(s) |
| 9 | ⚡ | Thunderstorm, with or without precipitation |

Codes 0, 1, 2: NOT PLOTTED

# INDEX

Absolute humidity, 131
Absolute temperature scale, 70
Absorption
    of solar radiation by atmosphere,
        56–58
    at various wavelengths by
        atmospheric constituents, 60–62
Absorptivity, 61
Acceleration of gravity, 71
Adiabat, 77, 140
    dry or unsaturated, 77
    wet or saturation, 140
Adiabatic cooling
    formation of clouds by, 138–139
    saturated rate of, 139–140
    unsaturated rate of, 76
Adiabatic lapse rate, 79
Adiabatic process, 76
    saturation, 139
    unsaturated, 76
Adiabatic rate of temperature change
    saturation, 139–140
    unsaturated, 76
Advection fog, 138
Agriculture,
    effect of on weather and climate,
        256
    special weather forecasts for, 231
Air
    composition of, 23–25
    density of, 71
Air masses, 163, 168–175
    arctic, 169, 171, 172–175
    classification of, 168–169

continental polar, 169, 171,
    172–175
continental tropical, 169
equatorial, 169, 170
maritime polar, 169, 171, 172–173
maritime tropical, 169, 171
source regions, 168
transformation of, 163, 170–174
weather in, 163, 174–175
Air-mass transformation, 170–175
Air-mass weather, 174–175
Airplane pilot weather reports
    (PIREPS), 216
Air pollution, 83–84, 238–254
    concentration of, meteorological
        factors affecting, 241–242
    control of, 252
    effect of, 250–252
    emission standards, 252
    potential, 245–249
Air Quality Act, 253
Air trajectories, 109
Albedo, 56, 57
Aleutian LOW, 125, 189
Altimeter, 74
Altocumulus clouds, 11, 12, 182
Altostratus clouds, 5, 7, 11
Ammonia, 26
Analogue method of weather
    forecasting, 209
Anemometer, 88
Angular momentum, 117–119
Angular velocity, 101
    of earth, 101–102

Anticyclones, 16, 87, 108
  in relation to air mass formation,
    168
  in relation to air pollution
    potential, 245–246
  subtropical, 123, 125, 130, 172,
    189, 196
  variation with height, 113
Anticyclonic flow, 108
Arctic air masses, 169, 171,
    172–175
Arctic front, 179–180
Arctic sea smoke, 137
Asiatic monsoon, 195
Assmann, R., 30, 31
Atmospheric circulation
  general, 115–125
  at low latitudes, 194–207
Atmospheric window, 62
Aurora australis, 40
Aurora borealis, 40
Average pressure distribution,
    125–130
Azores HIGH, 125

Baguios, 199
Baroclinic zones, 179, 191
Baroclinicity, 191
Barometer, 3, 68
Barotropic regions, 191
Bergeron process, 145–148
Bermuda HIGH, 125
Billow clouds, 182
Bjerknes, J., 178, 182
Black, Joseph, 25
Black body, 43
Black-body radiation, 44, 45
Bora, 98
Brandes, H. W., 211
Breeze, 86
  sea and land, 92–93, 104, 163

Calorie, 46, 47
Carbon dioxide, 25, 26
  absorption of radiation by, 61
  trend in concentration of, 258
Carbon monoxide, 26, 240, 251, 253
Celsius, Anders, 69

Celsius temperature scale, 69–70
Centigrade temperature scale, 69–70
Centrifugal force, 108
Centripetal acceleration, 107
cgs system of units, 71
Chinook, 98, 171
Circle of inertia, 104
Circulation of atmosphere
  general, 115–125
  at low latitudes, 194–207
Cirrocumulus, 11, 12
Cirrostratus clouds, 5, 7, 11
Cirrus clouds, 5, 6, 11
Clean Air Act (British), 250
Clean Air Act (U.S.), 253
Clean Air Amendments of 1970, 253
Climate
  and air masses, 169–175
  associated with wind systems, 123
  effect of human activities on,
    256–261
  of tropical regions, 195–203
Cloud classification, 9–11
Cloud drops, number and size of, 144
Clouds, 3ff, 170
  altocumulus, 11, 12, 182
  altostratus, 5, 7, 11, 183, 184
  cirrocumulus, 11, 12
  cirrostratus, 5, 7, 11, 183, 184
  cirrus, 5, 6, 11, 183
  convective, 3, 153, 163, 175
  cumulonimbus, 3, 5, 11, 155, 184
  cumulus, 3, 4, 11, 155, 162,
    171, 184
  lenticular, 17
  nimbostratus, 8, 11, 183
  stratocumulus, 11, 13, 196
  stratus, 11, 13, 172, 175, 196
  trade-wind cumulus, 172, 196
Cloud seeding, 148, 262–269
  randomized tests of, 263–266
Cold front, 8, 9, 180, 182
  relation of tornados to, 166
Cold front aloft, 187
Collision-coalescence process, 145
Collision efficiency, 145
Computer-produced charts, 218
Condensation, 135

Condensation nuclei, 136, 146
Conditional instability, 142
Constant-level balloons, 236
Continental polar air masses, 169,
      171, 172–175
Continental tropical air masses, 169
Contours, isobaric, 90, 189
Convection, 3, 154
Convective cells, 153, 155
Convective clouds, 3, 153, 163
Convective instability, 160–161
Convergence, 159–160, 165
Coriolis, G. G. de, 99
Coriolis force, 99, 100–104, 194, 196
Coriolis parameter, 103
Cumulonimbus clouds, 3, 5, 11, 155
Cumulus clouds, 3, 4, 11, 162
Cumulus congestus, 155
Cumulus humilis, 155
Curved flow, 106
Cycles, 209
Cyclogenesis, 188, 191
Cyclonic flow, 108
Cyclonic shear, 179
Cyclones, 16, 87, 108
   Bay of Bengal, 199, 205–207
   family of, 187–188
   life cycle of, 184–188
   models of, 177
   tropical, 87, 199–207
   wave, 8, 179, 181–184

Day, length of, 53
Density, 70–71
Depression, tropical, 200
Deserts, 123
Diamond dust, 150
Dishpan experiments, 120–121
Diurnal heating, 63–64, 82
Diurnal variation
   of pressure, 3
   of temperature, 63–64
Divergence, 161–162
Doldrums, 123
Donora, Pa., air-pollution disaster,
   250
Dosage of air pollution, 248–249
Dove, H. W., 177

Downdraft, 155
Drizzle, 149, 163
Drop size in clouds, 144
Dropsondes, 236
Dry adiabat, 77, 140
Dry adiabatic rate of cooling, 139
Dust devils, 153
Dynamic weather prediction, 208
Dyne, 69

Earth
   angular velocity of, 101–102
   distance from sun, 39
   effect of rotation of, 99–101
   orbit around sun, 39, 51
Easterlies
   polar, 123
   trade winds, 86, 99, 122–123,
      124, 194–195, 196–197
Easterly waves, 197–198
Effects of human activities on
      weather, 256–261
Electromagnetic radiation, 41–45
   black-body, 43, 44, 45
   global, 50
   laws of, 43–44
   spectrum, 42–43
Electromagnetic waves, speed of, 42
Emission standards (air pollution),
   253
Emissions of pollutants, 240
Energy, 42, 45–47
   chemical, 46
   of hurricanes, 203
   kinetic, 46
   potential, 46
   radiant, 46
   solar, 39–40
Entrainment, 158
Environmental Protection Agency,
   253
Equation of state, 68
Equatorial air mass, 170
Equinoxes, 52, 53
Espy, James P., 211
Evaporation, 134
Extended-range forecasting, 234–235
Extremes of temperature, 35–36

Eye of the storm (hurricane), 201,
  203–204
Eye wall, 201, 204

Fahrenheit, G. D., 70
Fahrenheit temperature scale, 70
Fair-weather cumulus, 155
Fallout, 260
Fall winds, 98
Fanning, 242
Fire-weather forecasts, 234
First GARP Global Experiment
  (FGGE), 236
Fitz-Roy, Robert, 177
Flash-flood warnings, 233
Flood forecasting, 233
Fly ash, 239
Foehn, 98, 171
Fog, 162, 163, 172, 175, 184, 196
  advection, 138
  artificial dissipation of, 267–268
  radiation, 138
  steam, 137
  upslope, 161
  warm frontal, 137
Forces in a fluid, 89–91
Forecasting. See Weather forecasting
Forecasts, accuracy of, 211–213
Fossil-fuel combustion, 238–239,
  256, 258
Foucault pendulum, 104
France, hail-suppression experiments
  in, 268
Freezing nuclei, 147
Friction, effect of, 109–110
Frontal inversion, 84, 181
Frontal slope, 181
Frontal waves, 181–184
Frontal weather, 184
Frontal zones, 179
Fronts, 8, 178–181
  arctic, 179–180
  cold, 8, 9, 166, 180, 182
  occluded, 185–187
  polar, 179
  warm, 8, 9, 84, 180, 182
Front, vertical cross section through,
  180

Frost
  prevention, 262
  warnings, 231
Fumigation, 242
Funnel cloud, 165–166

Gale, 86
Gamma rays, 43
GARP Atlantic Tropical Experiment
  (GATE), 236
Gas, 66
Gas constant, 69
Gas laws, 67, 68
Gay-Lussac, J. L., 29
General circulation, 115–125
General circulation experiments,
  120–122
Geostrophic balance, 105
Geostrophic wind, 104–106
Glaze, 151
Global Atmospheric Research
  Program (GARP), 235–236
Gradient, pressure, 91
Gradient wind, 108–109
Gradient wind speed, 109
Graupel, 150
Gravity, acceleration of, 71
Greenhouse effect, 61, 64, 260
Ground inversion, 83

Hadley, George, 99, 117
Hadley cells, 117
Hadley circulation, 117, 119
Hail, 148, 151, 165
Hailstone, 151–152
Hail suppression, 268
Haze, 23
Heat, 46–47
  latent, 139, 203
Helium, 25
Henry, Joseph, 211
Heterosphere, 27, 28
HIGH pressure centers. See
  Anticyclones
Homosphere, 27, 28
Horizontal convergence, 159–160,
  165

Horse latitudes, 123
Howard, Luke, 10
Humidity, 131–135
  absolute, 131
  relative, 133–135
  specific, 133
Hurricanes, 86, 87, 199–207
  modification experiments, 268–269
  structure of, 201, 203–204
  tracks of, 203
  warnings, 233
Hydrocarbons, 239, 240, 251, 253
Hydrostatic equation, 72
Hygrograph, 134
Hygroscopic nuclei, 137
Hygrothermograph, 134

Ice crystals, 147, 150
Icelandic LOW, 125, 189
Ice nuclei, 147
Ice pellets, 150
Ice storms, 151
Impaction of particulates, 260
India, precipitation augmentation
  experiments in, 267
Inertial period, 104
Infrared radiation, 42–43
Insolation, 50
  reduction of in cities, 258
Instability, 80–82
  conditional, 142
  convective, 160–161
International Council of Scientific
  Unions, 235
Intertropical convergence zone (ITC),
  123, 130, 195, 196
Inversion, 82–84, 169, 242
Ionosphere, 3, 27, 28
Irradiance, 47
Isobaric contours, 90, 189, 218
Isobaric process, 77
Isobars, 16, 90, 218
Isotherms, 34, 218
Israel, weather modification
  experiment in, 266
Italy, hail suppression experiments
  in, 268

Jeans, Sir James, 44
Jet stream, 87, 189–191

Kelvin, Lord (William Thompson),
  70, 135
Krypton, 25

Land breeze, 92–94
Langley, 47
Langley, Samuel P., 47
Lapse rate, 79
  adiabatic, 79
Latent heat
  of melting, 139
  as source of energy of hurricanes,
    203
  of sublimation, 139
  of vaporization, 139
Lateral mixing, 244
Lavoisier, Antoine L., 211
Layers of the atmosphere, 28
Lead, 251
Length of day, 53
Lenticular clouds, 171
Lifting condensation level, 141
Lifting condensation pressure, 141
Lifting condensation temperature,
  141
Light, wavelength of visible, 42–43
Lightning, experiments to decrease,
  268
Lofting, 243
London, England, air-pollution
  disaster, 250
London-type smog, 240
Long-wave radiation, 60–61
Looping, 242
Los Angeles-type smog, 240
LOW pressure centers. See Cyclones

Magnetic field, 28
Magnetosphere, 28
Map analysis, weather, 213, 217–218
Maritime polar air mass, 169, 171,
  172–175
Maritime tropical air mass, 169, 171,
  175
Mayow, John, 24

Mercury barometer, 68
Mesopause, 27, 28, 29, 30
Mesosphere, 27, 28, 29, 30, 32
Meteorograph, 31
Meteorological satellites, 17ff
Methane, 26
Micron, unit of length, 42
Millibar, unit of pressure, 69
Mist, 149
Mistral, 97
Mixing condensation level, 142
Mixing condensation pressure, 142
Mixing condensation temperature,
    142
Mixing height, 246
Mixing ratio, 133
mks system of units, 71
Moisture in the atmosphere, 24,
    131ff. *See also* Humidity
Momentum, 88
Momentum, angular, 117–119
Monsoon, 97, 163
    Asiatic, 195
Motion of air, 86ff
    effect of earth's rotation, 99–101
    equation of, 89, 221
    Newton's laws of, 88–89
    scales of, 87–88
Mountain and valley winds, 95
Mountain wave, 171

National Oceanic and Atmospheric
    Agency, 211
Neon, 25
Neutral equilibrium, 80
Newton, Isaac, 88
Newton's laws of motion, 88
Nimbostratus clouds, 8, 11
Nitrogen, 24, 25
    oxides of, 26, 253
Nitrogen dioxide, 239, 251
Nocturnal thunderstorms, 162
Northeast trades, 123
Nuclear power, 253–254
Nuclear reactors, 256
Nuclei
    condensation, 136, 146
    hygroscopic, 137

ice, 147
    sublimation, 147
Numerical weather prediction
    (NWP), 20–21, 208, 209,
    220–230

Observations
    airways weather, 216
    pilot balloon, 214
    radar, 217
    radiosonde and rawinsonde,
        214–216
    sferics, 217
    surface, 10, 213, 214
    upper-air, 10, 213–216
Occluded front, 185–187
    cold-front type, 187
    warm-front type, 187
Occlusion, 185
Occlusion process, 185–186
Orchard heaters, 262
Overseeding, 266
Oxides of nitrogen, 253
Oxygen, 24, 25
    absorption of radiation by, 57, 61
    dissociation of, 57
Ozone, 57, 251
    absorption of radiation by, 57
    formation in photochemical
        smog, 239
Ozonosphere, 27, 28

Pakistan, precipitation augmentation
    experiments in, 267
Parallelogram rule of vector
    addition, 89
Partial pressures, law of, 132
Particles in atmopshere, 24
    scattering of radiation by, 56
Particulate concentration, trend of,
    260
Particulate pollution, 240
Particulate removal processes, 260
Pendulum day, 104
Perihelion, 39
Periodicities, 209
Phase changes, 139

Photochemical smog, 239
Pilot balloon observations
   (PIBALS), 214, 216
Planck, Max, 44
Planck's radiation law, 44
Plotting models, weather map
   surface, 16
   upper air, 218
Polar air, 163, 169
Polar easterlies, 123
Polar front, 179
Polar front theory of cyclones, 178
Pollutants, maximum allowable
   concentration, 253
Pollution concentration, diurnal
   variation of, 247
Pollution-episode days, 249
Pollution sources, 240
   types of, 241
Potential temperature, 78
Precipitation, 142, 148–152
   in air masses, 163–165, 175
   artificial stimulation of, 263–267
   cloud seeding, 263–267
   drop growth by collision and
      coalescence, 145
   drop growth by three-phase
      (Bergeron) process, 145–148
   forecasts, 222–230
   forms of, 149–153
   at fronts, 182, 184
   in hurricanes, 201
Predictability, 209–210
Predictand, 230
Prediction. See Weather forecasting
Predictors (in statistical forecasting),
   230
Pressure, 3, 14, 66, 69, 89–90
   average sea-level over Northern
      Hemisphere, 126–127
   diurnal variation of, 3
   normal sea-level, 69
   reduction to sea-level, 74
   in Standard Atmosphere, 74
   vapor, 132, 135–136, 147
   variation in the horizontal, 90
      causes of, 92–93
   variation with height, 72

Pressure gradient, 91
   variation with height, 111
Pressure-gradient force, 90–91
Prevailing westerlies, 99, 123, 124
Priestley, Joseph, 25
Process curve, 79
Prognostic charts, 222, 231
Project Whitetop, 266
Psychrometer, asperated, 30, 31

Quantum theory, 44

Radar observations, 217
Radar storm detection, 233
Radiant flux density, 47
Radiation, 40ff
   absorption of 56–58, 60–62
   black-body, 43
   corpuscular, 40
   electromagnetic-wave, 41–45
   infrared, 42–43
   laws of, 43–44
   long-wave, 60–61
   Planck's law, 44
   reflection of, 56, 57
   scattering, 56
   short-wave, 60
   solar, 47–59
   spectrum, 43
   Stefan-Boltzman law of, 44
   terrestrial, 60–64
   ultraviolet, 42–43
   variation with latitude, 53–55, 116
   Wien's displacement law, 44
Radiation balance, 116
Radiation budget, 64
Radiation fog, 138
Radiation inversion, 83
Radiative equilibrium, 116
Radio waves, 42–43
   reflection by ionosphere, 3, 27
Radiosonde, 10, 31
   observations (RAOBS), 214–216
   station network, 216
Radius of curvature, 108
Rain, 150. See also Precipitation
   freezing, 151
Rain gauges, 149

Rainmaking, 263
Rainout, 260
Ramsay, Sir William, 25
Randomized cloud seeding
   experiments, 263–266
Rawinsonde, 10
   observations (RAWINS), 214, 216
Rayleigh, Lord, 25, 44
Reflectivity. *See* Albedo
Relative advection, 159
Relative humidity, 133–135
Representative point (on
   thermodynamic diagram), 77
Residence time of pollutants in
   atmosphere, 240
River-stage measurements, 233
Rocket observations, 32
Rossby, C. G., 189
Rossby waves, 189
Rotation of earth
   around vertical axis, 102
   effect on moving air, 99–104
   rate of, 101–102
Rutherford, Daniel, 25

Salt particles as condensation nuclei,
   137, 146
Santa Ana wind, 98, 171
Satellites, meteorological, 17–20, 217,
   235–236
Satellite photographs, 192, 202, 206,
   217
   identification of hurricanes on, 201
Saturation, 133
Saturation adiabat, 140
Saturation adiabatic process, 139–140
Saturation adiabatic rate of cooling,
   139–140
Saturation vapor pressure, 134
Scales of atmospheric motion, 87–88
Scattering, 56
Schaeffer, Vincent, 148, 262
Scheele, Carl William, 25
Sea breeze, 92–93, 104, 163
Sea-breeze front, 95–96
Sea-level pressure, 74
Seasonal forecasting, 235

Seeding of clouds. *See* Weather
   modification
Severe storm forecasting, 233–234
Shear of wind
   at fronts, 179
   in vertical, 111
Short-wave radiation, 60
Showers, 150, 163, 165
Siderial day, 102
Silver iodide, as ice nucleant for
   cloud seeding, 148, 262
Sky, color of, 2, 56
Sleet, 150
Slope of fronts, 181
Slope wind, 95
Small hail, 151
Smog, 239
   photochemical, 239
   Los Angeles-type, 240
   London-type, 240
Snow, 150
   albedo of, 57
   radiation by, 173
Snow crystals, 132
Snowfall, 150
Snowflakes, 150
Snow pellets, 150
Snow showers, 162
Snow survey, 233
Soft hail, 150
Solar constant, 48, 49
Solar radiation, 47–59
   absorption at earth's surface, 56, 59
   depletion by atmosphere, 56–59
   variation with latitude and season,
      50–56
Solar system, age of, 40
Solberg, H., 182
Solstices, 52, 53
Soot, 239
Sounding, 79
Sounding curve, 78–79
Southeast trades, 123
Source region, 168
Soviet Union, hail-suppression
   operations in, 268
Specific heat, 75
Specific humidity, 133

Specific volume, 69, 72
Spectrum
  absorption, 61–62
  electromagnetic, 43
Spiral cloud bands (of a hurricane),
    201, 203, 204
Squall line, 9, 163–165, 166
Stability, 79–82
  criteria for cloudy or saturated air, 142
  criteria for unsaturated air, 82
  as factor in air-pollution
      concentration, 242
  processes producing changes in,
      159–162
Stable equilibrium, 80
Standard Atmosphere, 28, 29, 30, 74
State, equation of, 68
Stationary front, 180
Statistical forecasting, 208, 230
Steam fog, 137
Steering of wave cyclones, 189
Stefan-Boltzmann law, 44
Storm (wind speed designation), 86.
    See also Cyclones, Hurricanes,
    Thunderstorm, Tropical storms
Storm surge, 204
Stratiform clouds, 5
Stratocumulus clouds, 11, 13, 196
Stratopause, 27, 28, 30, 30, 34
Stratosphere, 27, 28, 29, 30, 32
  discovery of, 31
Stratus clouds, 11, 13, 149, 163, 196
Streamlines, 106, 194
Sublimation nuclei, 147
Subsidence, 172, 245–246
Subtropical high-pressure belts, 123
Subtropical high-pressure centers,
    125, 130, 172, 189, 196
Subtropical inversion, 83
Sulfur dioxide, 26, 239, 240, 251
Sun
  distance from earth, 39
  energy emitted by, 39–40
Supercooled drops, 146
Supersaturation, 136
Surface weather observations, 213
Switzerland, hail-suppression
    experiments in, 268

Synoptic meteorology, 14
Synoptic observations, 213
Synoptic weather forecasting, 208
Synoptic weather maps, 217

Teisserenc de Bort, L. P., 30, 31
Temperature, 69
  adiabatic change of, 76
  average at earth's surface, 36–37
  change due to addition of heat, 75
  extremes (record high and low on
      earth), 35–36
  forecasting, 230–231
  increase due to increase of carbon
      dioxide, 261
  lapse rate, 79
  potential, 78
  trend of earth's average, 260
  urban heat island, 257
  variation with latitude and season
      at various heights, 32ff
  wet-bulb potential, 140, 141
Temperature inversion, 82–84
Temperature scales, 69–71
Terminal velocity of drops, 143–144
Terrestrial radiation, 60–64
Theodolite, 214
Thermal instability, 153
Thermal wind, 112–113
Thermals, 154
Thermistor, 215
Thermodynamic diagram, 77
Thermograph, 134
Thermometer, 30, 70
Thermosphere, 27, 28, 30
Thickness lines, 218
Thompson, Benjamin (Count
    Rumford), 46
Thompson, William, Lord Kelvin, 70,
    135
Three-phase process of rain
    formation, 145–148
Thunderstorm, 154, 155, 163–165,
    167
  air-mass, 163, 172
  cells, 155
  frontal, 183–184
  line, 163–165

Thunderstorm (cont'd)
  nocturnal, 162
  stages in development of, 155, 157
Tidal wave, 204
TIROS weather satellites, 17–19
Tornados, 9, 87, 165–167
  diameter of, 165
  diurnal variation of, 167
  locations of frequent occurrence,
    166
  paths of, 166
  pressure drop in, 166
  seasonal variation of, 166–167
  wind speed in, 165
Tornado warnings, 233–234
Torque, 118
Trade-wind cumulus, 172, 196
Trade-wind inversion, 83, 172, 196,
    246
Trade winds, 86, 99, 122–123, 124,
    194–195, 196–197
Transport of heat and momentum by
    vortices, 120
Tropical air masses, 169, 171, 175
Tropical cyclones, 87, 198–207
  classification of, 199–200
  structure of, 201
  tracks of, 203
Tropical depression, 200
Tropical disturbance, 200
Tropical maritime air, 163, 169, 171
Tropical storm, 200
Tropopause, 27, 28, 29, 30, 34–35
Troposphere, 27, 28, 30, 32
Turbidity, 259
Turbulence, 111
Turbulent mixing, 141, 158, 244–245
Typhoons, 87, 199

Ultraviolet radiation, 42–43
Units, systems of, 71
Unsaturated rate of adiabatic cooling,
    139
Unstable equilibrium, 80
Updraft, 155
Upper-level maps, 218
Upwelling, 197

Urban effects on weather, 257
Urban heat island, 257
U.S. National Meteorological Center
    (NMC), 218
U.S. National Weather Service, 211
U.S. Standard Atmosphere, 28, 29, 30,
    74

Valley-mountain wind, 95, 163
Van Allen belts, 28
Vapor lines, 140
Vapor pressure, 132
  equilibrium over drops, 135–136
  saturation over ice, 135, 147
  saturation over water, 134–135
Vectors, 89
Vertical structure of atmosphere, 26ff
Vertical variation of pressure systems,
    113–114
Viscous stress, 89, 110
Visibility, reduction of
  by air pollution, 251–252
  in cities, 258
Visible light, 42, 43
Volcanic eruptions, effect on
    turbidity, 259
Vonnegut, Bernard, 148, 262
Vortices, 87, 120, 123
  anticyclones, 16, 108
  cyclones, 16, 108
  dust devils, 153
  hurricanes, 87, 199–207
  tornados, 165
  transport of heat and momentum
    by, 120
  tropical cyclones, 87, 199–207

Warm front, 8, 9, 84, 180, 182
Warm frontal fog, 137
Warm sector, 182
Warnings
  flood, 233
  frost, 231
  hurricane, 233
  severe storm, 233
Washout, 260

Waterspouts, 167
Water vapor, 131
  absorption of radiation by, 61
Waves, 41
  frequency, 41, 42
  period, 41–42
  speed, 41, 42
  stability, 182
Wave clouds. *See* Lenticular clouds
Wave cyclone, 8, 179, 181–184
Wavelength, 41
Waves, 120
  in the easterlies, 197–198
  mountain lee, 171
  in the westerlies, 188–191
Weather
  air-mass, 174–175
  frontal, 184
  in hurricanes, 204
Weather forecasting, 17, 191–192,
    208–236
  accuracy of, 211–212
  analogue method, 209
  dynamic, 208
  extended range, 234–235
  numerical weather prediction
      (NWP), 208, 220–231
  statistical, 208, 230
  synoptic, 17, 208
  from weather maps, 17, 208
Weather map, 10ff
Weather-map analysis, 213, 217–218
Weather message code, 213
Weather modification, 148, 261–269
Weather observations. *See*
    Observations
Weather reconnaisance planes, 217
Westerlies, 124
Westerly waves, 189
Wet adiabat, 140
Wet-bulb potential temperature, 140,
    141
Wet-bulb thermometer, 30–31
Wien, Wilhelm, 44
Wien's displacement law, 44
Willy-willies, 199
Wilson, C. T. R., 135

Wind, 86
  anticyclonic, 108
  average distribution over globe,
      122–123
  cyclonic, 108
  distribution in low latitudes,
      194–195
  effect of friction on, 109–111
  geostrophic, 104–106
  gradient, 108–109
  hurricane, 86, 199
  mountain and valley, 95
  surface, deviation from geostrophic,
      109–111
  thermal, 112–113
  in tornados, 165
  in tropical storms, 199–201
  variation with height, 111
Windbreaks, 262
Wind machines, 262
Wind shear, 111
Wind speed
  diurnal variation of, 246–247
  effect on air-pollution
      concentration, 244
Windstorms, 165
Wind systems, 122–125
  doldrums, 123
  intertropical convergence zone
      (ITC), 123, 130, 195, 196
  monsoon, 97, 163, 165
  polar easterlies, 123
  prevailing westerlies, 99, 123, 124
  trade winds, 86, 99, 122–123, 124,
      194–195, 196–197
Wind vane, 88
World Meteorological Centers, 217
World Meteorological Organization
    (WMO), 213, 235
World Weather Watch (WWW),
    235–236

Xenon, 25
X-rays, 43

Zonal flow, reason for, 115
Zonal wind systems. *See* Wind
    systems